Quantitative and Dynamic
Plant Ecology

To Ellen and her hearty
disrespect of too much jargon

Quantitative and Dynamic Plant Ecology

Second edition

Kenneth A. Kershaw

Professor of Biology

McMaster University, Ontario

American Elsevier Publishing Company, Inc.

New York

American Elsevier Publishing Company, Inc.
52 Vanderbilt Avenue, New York, N.Y. 10017

First published in Great Britain by
Edward Arnold (Publishers) Ltd.

Cloth edition ISBN: 0-444-19554-8
Paper edition ISBN: 0-444-19545-9

Library of Congress Catalog Card Number: 73-10275

Second printing 1975

Printed in Great Britain by
William Clowes & Sons, Limited
London, Beccles and Colchester

Preface to the First Edition

THE study of Ecology now necessitates a knowledge of many sciences in addition to Botany and Zoology and with the rapid development of statistics the student of Ecology is often faced with the problem of using and understanding the simpler statistical procedures which are in common use.

The aim of this book is to present the general theoretical concepts that have been formulated in relation to plant ecology together with the approaches and statistical methods that have been used in developing these concepts. The early work in plant ecology was, of necessity, largely concerned with the description of vegetation in relation to the environment. This branch of ecology is still very actively pursued and will continue to be of importance until all vegetation is described on a world-wide basis. From this early descriptive ecology came a realization of the difficulties of sampling vegetation and its dynamic nature, and consequently sampling measures were developed and tested and different measures of the abundance of a species devised. The concept of plant succession was also developed from this early descriptive work and pointed the way to many of the concepts in use today. With the basic idea of a dynamic vegetation accepted, further examination in still greater detail outlined some of the processes involved and demonstrated the close inter-relationships that exist between species and between individuals of the same species. On reaching this level of development the necessity for sensitive sampling measures and statistical procedures to extract from a mass of data the relevant information became of paramount importance.

This book follows the general sequence of development outlined above. The early chapters are concerned with the methods of describing vegetation and the problem of sampling. The subsequent chapters deal with the fundamental concepts in use today. These include succession, cyclic change, plant interactions and the non-random distribution of individuals of a species in vegetation. The final chapters outline the controversy over the validity of the plant community as a distinct and recognizable entity and the methods that have been employed in this branch of ecology.

No detailed description of vegetation types or account of environmental factors and their measurement is attempted—both these aspects of ecology are now so extensive as to merit a book to themselves.

Finally, it gives me considerable pleasure to thank Dr. N. Waloff of the Department of Zoology, Imperial College, for her reading of the draft manuscripts and her many valuable criticisms and suggestions. I also owe a considerable debt to Dr. A. J. Rutter of this Department for the many hours he has spent in criticizing and discussing various sections of this book.

Kenneth A. Kershaw

Imperial College
　London, S.W.7
1964

Preface to the Second Edition

SINCE the publication of the first edition of this book there has been a tremendous increase of interest in classification and ordination of vegetational samples or units. As a result the original chapter on classification and ordination methods has been extensively revised and developed as two separate chapters. The emphasis is towards an understanding of the principles involved rather than an attempt at explaining the mathematical complexities. Two other developments in plant ecology, population studies and computer methods, are also well established areas of ecology necessitating a numerical approach. The former has been a gradual development over the past ten years and forms a new chapter (chapter 6). The latter development has been more explosive and has also necessitated a chapter on computer methods (chapter 11). This again is not intended as a detailed introduction of programming method but an attempt at a general outline of some of the more powerful features of the present generation computer. Chapter 12 covers the principles of computer simulation but avoids the mathematical treatments often mistakenly identified with simulation studies. Finally new data has been incorporated into chapters 4 and 8.

I am indebted to Dr. G. P. Harris for his criticisms of the drafts of the new chapters and his helpful suggestions. I am grateful to him also together with Dr. J. N. A. Lott for their permission to use some of their data on hysteresis effects in algal populations prior to its publication. I am most grateful to Mrs. Janet Edmonds for her help in tracing references, and her invaluable assistance in the preparation of the manuscript for the press, and to Miss Judy Street, who has typed successive drafts.

Thanks are due to the Literary Executor of the late Sir Ronald A. Fisher, F.R.S., Dr. Frank Yates, F.R.S. and to Longman Group Ltd., London, for permission to reprint the tables on pages 274-285 which are from *Statistical Tables for Biological, Agricultural and Medical Research* (formerly published by Oliver and Boyd, Edinburgh).

Kenneth A. Kershaw

McMaster University
  Hamilton, Ontario
1973

Contents

1 The Description of Vegetation

A CONSIDERABLE proportion of all ecological work in the past and to a considerable extent at the present time, has been directed towards the description of vegetation. The object of such description is to enable people other than the observer to build a mental picture of an area and its vegetation and to allow the comparison and ultimate classification of different units of vegetation. Before any serious or detailed work can be commenced in an area it is necessary to know what species are present, what is their distribution and what is the relative degree of abundance of each species. Additional information on the morphology of the species characteristic of the area and a general account of the environment are also essential. Thus the *Floristic* composition, simply expressed as a list of species, *Life form* composition and *Structure* of vegetation are a necessary basis of all ecological work; this approach to ecology has dominated the subject until quite recently and will continue to be prominent until an adequate description of all vegetation both on a regional and on a world-wide basis is completed.

LIFE FORM

The best-known description and classification of life forms, and the use of life form to construct a *biological spectrum* are due to Raunkiaer (1909, 1928, 1934). He arranged the life forms of species in a natural series in which the main criterion was the height of perennating buds, and he showed that this criterion appeared to reflect adaptation to climate. He based this series on the assumption that the original environment was considerably more moist and uniformly hot than it is now, and accordingly the most primitive life form is that which dominates tropical vegetation at the present time. Conversely the more highly evolved species are found in the colder areas of the world. Raunkiaer established the following types (see also Fig. 1.1):

1. *Phanerophytes*

   The perennating buds or shoot apices borne on aerial shoots.
   (*a*) EVERGREEN PHANEROPHYTES without bud scales (tropical tree species)
   (*b*) EVERGREEN PHANEROPHYTES with bud scales

(*c*) DECIDUOUS PHANEROPHYTES with bud scales

(Each of the above groups can be sub-divided into Nanophanerophytes (<2 m in height), Microphanerophytes (2–8 m), Mesophanerophytes (8–30 m) and Megaphanerophytes (>30 m in height). Two further groups are usually included—epiphytic Phanerophytes and stem-succulent Phanerophytes.)

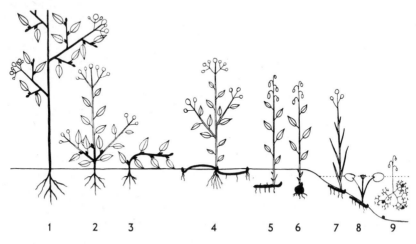

Fig. 1.1. The relative positions of the perennating parts of four life forms. (1) Phanerophytes, (2–3) Chamaephytes, (4) Hemicryptophytes and (5–9) Cryptophytes. The persistent axes and surviving buds are shown in black. (From Raunkiaer 1934; courtesy of Clarendon Press, Oxford.)

2. *Chamaephytes*

Perennating buds or shoot apices borne very close to the ground.

(*a*) SUFFRUTICOSE CHAMAEPHYTES. Characterized by more or less erect aerial shoots which die away in part at the onset of an unfavourable season of the year. Perennating buds arise on the lower portions of the erect stems where they are less exposed to the environment. Very characteristic of many parts of the Mediterranean.

(*b*) PASSIVE CHAMAEPHYTES. Similar to the last group but at the onset of adverse conditions the weakened erect axes fall over and buds arise along the horizontal stems where at ground level they obtain some protection from the environment; e.g. *Stellaria holostea.*

(*c*) ACTIVE CHAMAEPHYTES. The vegetative shoots are persistently orientated along the ground usually rooting along their length; e.g. *Lysimachia nummularia, Empetrum nigrum.*

(*d*) CUSHION PLANTS. Basically a reduced and very compacted form of the last group.

3. Hemicryptophytes

Perennating buds at ground level, all above ground parts dying back at the onset of unfavourable conditions. Stolons may or may not be present.

(a) PROTO HEMICRYPTOPHYTES. Lowermost leaves on stem less perfectly developed than the upper ones, the perennating buds arise at ground level; e.g. *Rubus idaeus*.

(b) PARTIAL ROSETTE PLANTS. The best developed leaves form a rosette at the base of the aerial shoot, but some leaves are present on the aerial stems; e.g. *Ajuga reptans, Achillea millefolium*, etc.

(c) ROSETTE PLANTS. The leaves are restricted to a rosette at the base of the aerial shoot; e.g. *Bellis perennis, Taraxacum officinale, Hypochaeris radicata*, etc.

4. Cryptophytes

Perennating buds below ground level or submerged in water.

(a) GEOPHYTES. Rhizome, bulb or tuber geophytes, over-wintering by food stores under ground from which arise the buds to produce the next season's aerial shoots.

(b) HELOPHYTES. Those plants which have their perennating organs in soil or mud below water-level with aerial shoots above water-level; e.g. *Typha, Alisma*, etc.

(c) HYDROPHYTES. Those plants with their perennating buds under water, and with their leaves submerged or floating. The buds may occur on rhizomes as in *Nuphar, Nymphaea*, etc. or winter buds may become detached from the plant and sink to the bottom of the water and there survive the unfavourable season (*Potamogeton obtusifolius, P. pusillus*, etc.).

5. Therophytes

Annual species which complete a life history from seed to seed during the favourable season of the year. Their life span can be as short as a few weeks, and they are characteristic of desert regions and cultivated soil where the interference of man protects them from their (slower growing) natural competitors.

Using the life form as a basis of comparison it is possible to compare areas which are widely separated geographically and do not contain species common to both (the normal basis of comparison of areas). The life form is assumed to have evolved in direct response to the climatic environment and accordingly the proportion of life forms in an area will give a good indication of its climatic zone. Thus the biological spectrum is a useful measure for comparison on a geographical scale and shows the number of species within each life form group (Table 1.1).

Table 1.1   The percentage representation of different life forms in the flora of Denmark (1084 species) and St. Thomas and St. Jan, West Indies (904 species). (From Raunkiaer 1934; courtesy of Oxford Clarendon Press)

	Denmark	St. Thomas and St. Jan
Mega and Mesophanerophytes (MM)	1	5
Microphanerophytes (M)	3	23
Nanophanerophytes (N)	3	30
Epiphytes (E)	0	1
Stem succulents (S)	0	2
Chamaephytes (Ch)	3	12
Hemicryptophytes (H)	50	9
Geophytes (G)	11	3
Hydrophytes and Helophytes (HH)	11	1
Therophytes (Th)	18	14

Raunkiaer gives several interesting examples from which he concludes that the zonation of vegetation progresses from a Phanerophytic type in the tropical zone to Therophytic in the sub-tropical desert areas, Hemicryptophytic in the cold temperate and finally a Chamaephytic type in the cold zone (Table 1.2). Similar trends can be seen with increase in altitude which

Table 1.2   The biological spectrum for different geographical zones. (From Raunkiaer 1934; courtesy of Oxford Clarendon Press)

Area	Percentage distribution of the species among the life forms (Key to symbols in Table 1.1)										
	S	E	MM	M	N	Ch	H	G	HH	Th	
Seychelles	1	3	10	23	24	6	12	3	2	16	Tropical forest
St. Thomas and St. Jan.	2	1	5	23	30	12	9	3	1	14	Tropical forest
Ellesmereland	—	—	—	—	—	23·5	65·5	8	3	—	Tundra
Baffinland	—	—	—	—	1	30	51	13	3	2	Tundra
Death Valley, California	3	—	—	2	21	7	18	2	5	42	Desert
Libyan Desert	—	—	—	3	9	21	20	4	1	42	Desert

reflects the increasing severity of the climate. Thus there are more Chamaephytes and correspondingly fewer Hemicryptophytes, with increase in altitude in the Alps (Table 1.3). In general the life form spectrum is not of great significance in comparing adjacent areas of vegetation but is of value in the realm of 'Plant Geography'.

Table 1.3 The biological spectrum at different altitudes in the Alps. (From Raunkiaer 1934; courtesy of Oxford Clarendon Press)

The nival region of the Alps	Percentage distribution of the species among the life forms (Key to symbols in Table 1.1)								
	S	E	MM	M	N	Ch	H	G	HH	Th
Above 3600 m						67	33			
3300–3600 m						58	42			
3150–3300 m						58	42			
3000–3150 m						52·5	45			2·5
2850–3000 m						33	62	2		3
2700–2850 m						33	61	3		3
2550–2700 m					0·5	28	65·5	2		4
2400–2550 m					1	24	67	4	0·3	4

STRUCTURE

The structure of vegetation is defined by three components: the vertical arrangement of species, i.e. the stratification of the vegetation; the horizontal arrangement of species, i.e. the spatial distribution of individuals; and, finally, the abundance of each species. The latter component of vegetation can be expressed in several ways, ranging from a direct count of the number of individuals in an area (density) to the dry weight of vegetable material produced in a given area (yield).

STRATIFICATION OF VEGETATION

A casual inspection of a woodland will demonstrate the marked stratification of the vegetation ranging from two layers (a herb layer at ground level and the canopy of the tree layer) to a more complicated arrangement with a herb layer, several shrub and sapling layers, and the overall canopy of the tree layer. In tropical forest the stratification of the vegetation has been long recognized as one of its characteristic features, though the variation of height within any one stratum has led to some argument as to the distinctness of the various strata recognized by different workers. The same criteria of layering can be applied to herbaceous vegetation and detailed description of the stratification of vegetation forms an important part in the understanding and recording of its structure. The method of description of the stratification of forest vegetation is due to Davis and Richards (1933–4) who constructed to scale, profile diagrams taken from narrow sample strips of tropical forest. (The concept of stratification is of obvious importance in any consideration of tropical forest and accordingly most of the published work on this topic relates to this type of vegetation.) The method consists of laying out a narrow rectangular sample plot, of required length and width (usually not less than 60 m in length, and 8 m has been found to be a convenient width). All vegetation lower than a chosen arbitrary height is cleared away, the remaining trees and shrubs being carefully mapped and the diameters of the trunks noted. The total height, height to first large branch, lower limit of crown and width of crown are measured. To obtain these measures it is often necessary to fell all the trees in the strip, commencing with the smallest trees so as to avoid the crushing of these smaller trees by the larger. From the measurements obtained an accurate scaled profile diagram can be obtained showing the spatial relation of the different species to each other both in a horizontal sense as well as a vertical sense.

Tropical forest is usually divided into five layers *A*, *B*, *C*, *D* and *E* (see Richards 1952). The *A* layer is composed of the tallest trees present in the area (usually termed 'emergents'), the *B* and *C* layers form two strata of tree species, *D* the shrub layer and *E*, the lowest stratum, the ground or field layer. Two examples of profile diagrams are given below (Figs. 1.2 and 1.3) taken from Richards (1936) and Beard (1949) and illustrate the

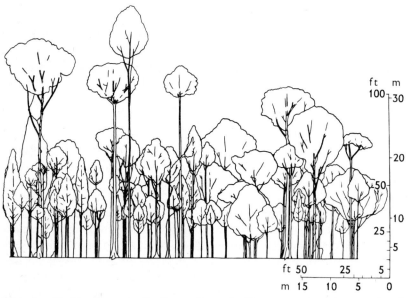

Fig. 1.2. Profile diagram of primary forest in Borneo representing a strip of forest 61 m long and 7·6 m wide. (From Richards, 1936; courtesy of *J. Ecol.*)

differences between primary mixed Dipterocarp forest in Borneo, having a distinct though discontinuous *A* layer with a continuous *B* and *C* layer and a single-dominant forest (*Dacryodes–Sloanes* association) in Dominica,

Fig. 1.3. Profile diagram of primary forest in the West Indies representing a strip of forest 7·6 m wide. (From Beard 1949; courtesy of *Oxf. For. Mem.*)

British West Indies, which has a well marked continuous *A* layer and clear *B* and *C* layers.

Profile diagrams can also be usefully employed in vegetation of lower stature to illustrate the relationship between topography and the distribution of individuals of a species, or the distribution of root systems in relation to the topography and drainage of an area. Thus, the distribution of vegetation in relation to hummocks in calcareous mire in Teesdale (Pigott 1956) is very conveniently expressed as a diagram (Fig. 1.4).

A simultaneous approach to stratification of vegetation, abundance and to spatial distribution of each species can be made in a way suggested by Christian and Perry (1953) who developed a method especially suitable for large-scale surveys. Letters and figures are assigned to the tree, the shrub and ground stories, to their component layers and to the density of each. Thus, A_1 is used for low trees, A_2 for medium sized trees and A_3 for tall trees. Similarly, B_1, B_2, B_3 and C_1, C_2, C_3 are used for low, mid-height and tall shrubs and herbs respectively. Heights can be recorded for each stratum as mean values. Thus, A_3^{15} would represent a tall tree 15 m in height, etc. Density is similarly treated by prefixing x, y or z for dense, average or sparse and xx or zz for very dense and very sparse respectively. Thus a complex community with two tree layers, three shrub layers and two grass layers at varying densities would be written as

$$A_3^{60}z, \; A_2^{30}y, \; B_3^{8}z, \; B_2^{3}x, \; B_1^{1}x, \; C_2^{1}zz, \; C_1^{3}zz.$$

This method has been used in a series of vegetational surveys in northern Australia and offers a good shorthand method of describing the physiognomy of the vegetation, the more abundant species of any layer being

Fig. 1.4. Profile diagram through the hummock complex at the upper margin of a small calcareous marsh. The depth of the humus-stained part of the hummocks is indicated by stippling. The lower diagram is a direct continuation along the same line as the upper. (From Pigott 1956; courtesy of *J. Ecol.*)

appended to the formula. It is clear that the height groups are quite arbitrary and that the errors associated with such visual assessments should be controlled by collaboration between different field workers. Comparisons of descriptions of the same stand made by different workers would reduce the errors to a minimal level, and accordingly increase the value of the method.

HORIZONTAL DISTRIBUTION—MAPPING TECHNIQUES

The detailed mapping of vegetation is obviously a very laborious and time-consuming operation and its use is normally reserved for the detailed study of small areas of vegetation made over a period of years. The mapped areas are in the form of 'permanent quadrats'. Permanent quadrats are established and carefully marked out in such a manner as to enable the exact position to be located again in subsequent years. Each individual plant, or shoot, or clump of shoots is recorded accurately. This enables a composite picture of change in vegetation with time to be built up. It is advisable to sub-divide the permanent quadrat to facilitate the positioning of individuals within the area, and similarly it is more convenient to record the actual positions of the individual plants on 'ruled square' paper. The relationship between *Festuca ovina* and *Hieracium pilosella* in grassland after the removal of rabbit grazing has been investigated by Watt (1962) who gives a series of permanent quadrats recorded at intervals from 1946 onwards (Fig. 1.5). The permanent quadrats show quite clearly the change over the years of the relative abundance of the two species.

In general, the use of both permanent quadrats and the mapping of vegetation, as descriptive techniques should be limited to those problems where the vegetational change is likely to be fairly rapid and at the same time very marked. Where vegetational change is less marked more refined techniques are necessary (see p. 14).

THE SUBJECTIVE ASSESSMENT OF ABUNDANCE

Frequency symbols

The simplest and most rapid way of describing vegetation is to list the species present within a sample area and to attach to each species a sub-jective assessment of its abundance. This method of approach has been used by numerous workers and several convenient classes of plant abundance have been employed. One of the most widespread of the subjective ratings used by British ecologists employs five classes of abundance, Dominant, Abundant, Frequent, Occasional and Rare with prefixes 'very' and 'locally' to qualify distribution and abundance where necessary. Thus

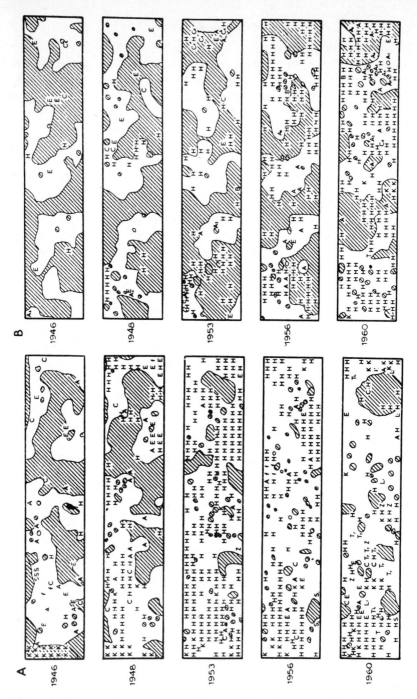

Fig. 1.5. The use of permanent quadrats to show the change of abundance of *Festuca* (shaded) and *Hieracium* (H) in enclosed (A) and control (B) plots. (From Watt 1962; courtesy of *J. Ecol.*)

a species of clumped habit, scattered over the area, would be referred to as locally frequent. Further classes can be used if thought necessary but it becomes increasingly difficult to distinguish between adjacent abundance classes as the total number of such classes increases. This in fact is one of the great weaknesses of this method of assessing the relative abundance of a species, and it is clear that other factors will often unknowingly influence the judgement of an observer. Small species tend to be rated lower than conspicuous ones, and similarly a species in full flower receives a higher rating than one which was in a vegetative state at the time of sampling.

Again, species that form obvious clumps are more conspicuous than species that are more evenly scattered and would be preferentially classed in an abundance scale. Thus such visual assessment of species abundance in an area includes a large unconscious error of judgement reflecting size, pattern of distribution and state of flowering of the species under consideration. Further the assessment is not constant from person to person, and a species that is referred to as 'Frequent' by one field worker may well be classed as 'Abundant' or 'Occasional' by others. This problem has been discussed by Hope-Simpson (1940) who investigated the causes of error in subjective assessment of abundance, including such effects as annual fluctuation, seasonal change and choice of sampling site. These latter errors are common to most measures of plant abundance and do not concern us here. However, comparisons of the data, derived from 'ordinary' and 'specially careful' recordings, illustrate the inconsistency of subjective methods. It is possible to compare the data if one establishes 'agreement classes' rated as 'Exact', 'Approximate', 'Intermediate' agreement or complete 'Disagreement'

Table 1.4 Comparison by agreement classes expressed as a percentage of the total number of species
(From Hope-Simpson 1940; courtesy of *J. Ecol.*)

	Comparison of two independent samples from the same site	Comparison of ordinary and specially careful recording
Exact agreement	44	36
Approximate agreement	8	12
Intermediate	36	38
Disagreement	12	14

The data shows clearly that subjective assessment of plant abundance contains a very large error and will, at the most, only give an approximate indication of abundance. Accordingly the use of frequency symbols should be strictly limited to initial description, and the degree of error present always borne in mind. A further drawback to the method lies in the rather unfortunate choice of the word 'Dominant' as the class representing the most abundant species, since in addition to the meaning here (the most abundant in a physiognomic sense) it is often confused with dominant in a sociological sense, meaning that species which exerts the most influence on the other species of the community. A species may be dominant in both senses of the word, but this need not necessarily be the case. For example Pearson (1942) has shown that regeneration of *Pinus ponderosa* depends on the type of ground vegetation. No regeneration occurs where the ground vegetation is *Festuca arizonica*, a species which grows during both the dry and the wet season, and can successfully compete with pine seedlings and prevent their establishment. On the other hand there is active regeneration when the ground vegetation is *Muhlenbergia montana* which only grows during the 'wet' season when water is plentiful for both species. In a sociological sense the dominant species is *Festuca arizonica*, whereas the immediate visual impression would put *Pinus* as the dominant. In many instances it is legitimate to equate the two definitions of 'dominant' but this is by no means always justified as the above example demonstrates. A further confusion that may arise is by the use of the word dominant in forestry literature as applied to a forest canopy, where those trees whose crowns are more than half exposed to full illumination are often referred to as 'dominants' (see Richards *et al.* 1940). It is usual that a dominant species of tree in the latter sense is one which is dominant in a sociological sense (one exception is given above), and thus it is recommended that 'dominant' as a frequency symbol be rejected—it can be very readily replaced by the term 'very abundant', which to all intents and purposes is synonymous.

Braun-Blanquet's system of rating

A more elaborate and better defined approach has been suggested by Braun-Blanquet (1927) who proposed what is now a recognized procedure for describing a stand of vegetation (see also Poore 1955). Two scales are used, one combining the number and cover of a species, and the second giving a measure of the grouping:

> $+$ = sparsely or very sparsely present; cover very small
> 1 = plentiful but of small cover value
> 2 = very numerous, or covering at least $\frac{1}{20}$ of the area
> 3 = any number of individuals covering $\frac{1}{4}$ to $\frac{1}{2}$ of the area
> 4 = any number of individuals covering $\frac{1}{2}$ to $\frac{3}{4}$ of the area
> 5 = covering more than $\frac{3}{4}$ of the area
> Soc. 1 = growing singly, isolated individuals
> Soc. 2 = grouped or tufted

Soc. 3 = in small patches or cushions
Soc. 4 = in small colonies, in extensive patches, or forming carpets
Soc. 5 = in pure populations

The assessment of plant abundance and distribution is obtained from a series of adjacent quadrats of increasing size which are laid down so as to continuously increase the sample area (Fig. 1.6). An exact list of all plant

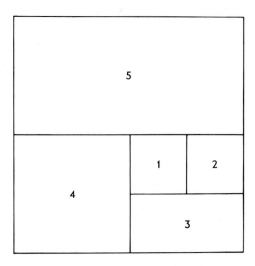

Fig. 1.6. The systematic arrangement of a series of quadrats to successively increase the size of the sample area. (From Poore 1955; courtesy of *J. Ecol.*)

species including lichens and bryophytes, with an attendant visual estimate of their sociability and abundance, is taken from the sample area. The measure of plant abundance used here is a definite improvement on the more traditional approach outlined previously. The scale combines, in part, abundance (number) with coverage of the area. It separates these to some extent from pattern of distribution and plant size which bias to an undefined extent the earlier abundance scale used. Accordingly the Braun-Blanquet scale has much to recommend it for a rapid survey of large areas of vegetation, since with practice it is possible for an individual recorder to minimize error from site to site. However, unconscious errors due to bias of different recorders will always be present and the extent to which they detract from the value of the method will remain debatable.

The Braun-Blanquet scale has several modifications and one of the more widely used scales, i.e. the Domin scale, is set out in Table 1.5.

The increased number of divisions enables a more detailed assessment to be made of plant coverage and field workers claim a high degree of accuracy can be obtained with sufficient experience. The Domin scale has

Table 1.5

	Domin scale	Braun-Blanquet
Cover about 100 per cent	10 ⎫	5
Cover > 75 per cent	9 ⎬	
Cover 50–75 per cent	8	4
Cover 33–50 per cent	7 ⎫	4
Cover 25–33 per cent	6 ⎬	
Abundant, cover about 20 per cent	5 ⎫	
Abundant, cover about 5 per cent	4 ⎬	2
Scattered, cover small	3 ⎫	
Very scattered, cover small	2 ⎬	1
Scarce, cover small	1 ⎭	
Isolated, cover small	x	x

the additional advantage in that it can be directly converted into the Braun-Blanquet scale wherever necessary and enable direct comparison of different sets of data.

QUANTITATIVE ASSESSMENT OF ABUNDANCE

Since the realization that a large degree of error is inherent in subjective evaluation of abundance, ecologists have become increasingly conscious of the necessity of using quantitative measures to describe vegetation. The question as to which abundance measure should be used is often affected by the time it takes to obtain a sample of the population of adequate size, and it is true to say that the better quantitative measures available to ecologists are certainly but unavoidably time consuming.

Density

This is a count of the number of individuals within an area. It is usual to count the number of individuals within a series of randomly distributed quadrats, calculating the average number of individuals relative to the size of quadrat used, from the total sample. The method is accurate, allows direct comparison of different areas and different species and is an absolute measure of the abundance of a plant. The disadvantage of the method is usually associated with the time involved in counting what, potentially, can be very large numbers of individuals. This can in part be circumvented by counting within small quadrats, but there always remains those cases where the 'individual' cannot be picked out with any great degree of certainty, e.g. *Vaccinium myrtillus*, many grasses, sedges, etc. In some instances a convenient unit of the vegetation can be used instead of the complete individual; thus, individual tillers, flowering culms, erect shoots, or, indeed, whole clumps could be a suitable reflection of the density of a species. When these approaches fail, as for example with a species such as *Festuca rubra*, it is necessary to find an alternative approach.

Cover

This is defined as the proportion of the ground occupied by perpendicular projection on to it of the aerial parts of individuals of the species under consideration (Greig-Smith 1957). It is in fact an estimation of the area covered by given species, usually expressed as a percentage of the total area and estimated from a number of sample points. The cover value can be obtained either by a visual estimate, a certain percentage of the total area of a quadrat being covered by a given species, or measured by taking a number of points from the sample area and determining at those points which species if any is covering the surface of the ground. Visual estimates are obviously subject to the same personal bias as has been discussed already in relation to frequency symbols (p. 11) and should be avoided where possible, despite their relative speed and ease of use. Measures of cover are conveniently obtained using a frame of pins that can be adjusted to the height of the vegetation. The pins are lowered one at a time, and the species touched by each pin in turn recorded. The final number of 'hits' from a number of sample 'frames' is then expressed as a percentage of the total number of pins. It should be noted that the total percentage cover for all species in an area will nearly always exceed 100 per cent since in closed vegetation leaves of several different species frequently overlap one another at a particular point.

Table 1.6 The dependence of estimated cover on the diameter of pin (From Goodall 1952a, courtesy of *Austr. J. Sci. Res.*)

Locality	Species	Number of points	Estimated cover using pin different diameters (mm)		
			0	1·84	4·75
Seaford	*Ammophila arenaria*	200	39·0	66·5	71·0
	Ammophila arenaria	200	60·5	74·0	82·0
Black Rock	*Ehrharta erecta*	200	74·5	87·0	93·5
Sorrento	*Lepidosperma concavum*	200	19·5	22·0	27·5
	Spinifex hirsutus	200	35·0	48·5	61·0
Carlton	*Fumaria officinalis*	200	20·5	31·5	30·0
	Ehrharta longiflora	200	24·5	25·5	37·5
	No contact	200	53·0	42·5	38·5
	Lolium perenne	200	65·0	85·5	82·5

The main error involved in this measure of plant abundance (apart from small personal errors which are inevitably present and are only of any consequence in tall vegetation) is the exaggeration of the estimate of percentage cover when a large diameter pin is used (Goodall 1952a). Goodall gives the data on the effect of pin diameter in Table 1.6.

For the majority of the species there is a considerable difference which is statistically significant, between the percentage cover estimate obtained by using an optical cross-wire apparatus sighted on to the vegetation (0 mm pin diameter) and the percentage cover reading given by the pins. This is simply due to the area of the tip of the pin being finite instead of zero. If absolute measures of percentage cover are required it is important to use an optical method, which will give a reliable measure of the true cover of a species. Conversely if data is being collected for purposes of comparison, providing the pin diameter is kept constant from site to site the exaggeration of the percentage cover is not of great importance except where the species tends to have a very high cover value. In the latter instance all areas will show a 100 per cent cover value whereas in fact the true value may range from 70 to 80 per cent in the different sites. If such very abundant species are to be included in a composite study it is necessary to use an optical method.

Percentage cover is a good measure of plant abundance and is very widely used in grassland where it is impossible to define where one individual starts and another ends. The main disadvantage of the method is the rather tedious and slow sampling involved. This can make the measure impractical in a large-scale survey for example. The difficulty in accurately observing the point of the needle in tall vegetation does lead to some error and there is a definite limit to the practicability of the method. The measure can be modified to take into account the stratified nature of the ground layer. The number of times a given species is encountered by a pin is recorded, rather than just the one hit noted in the normal measure of cover. Such a measure, often termed 'cover repetition', gives an estimate of the 'bulk' of plant materal and can thus be related to yield (see p. 20). Personal error in the measure of cover repetition tends to be rather high, and if an investigation calls for such a detailed approach it is probably more satisfactory to measure directly the actual weight of plant material in a sample area.

Frequency

The frequency of a species is a measure of the chance of finding it with any one throw of a quadrat in a given area. Thus, if a species has a frequency of 10 per cent then it should occur once in every ten quadrats examined. The measure is obtained very simply by noting whether a species is present or not in a series of randomly placed quadrats. For example, if 86 quadrats out of a total of 200 contain a given species the frequency is $86/200 \times 100 = 43$ per cent. If a value for the percentage frequency of a species within a single quadrat is required it can be simply obtained by sub-dividing the quadrat into a number of smaller units, then counting the number of sub-divisions within which the species occurs and expressing this as a percentage of the total number of sub-divisions. The results from a series of random quadrats can be finally averaged giving a value often termed local frequency.

Two convenient forms of frequency are used, *shoot frequency* and *rooted frequency* (Greig-Smith 1957). The former is obtained by including as 'present' all parts of the foliage that overlaps into a quadrat; the latter measure only includes a species as present when it is actually rooted within the area of the quadrat. The only advantage of frequency as a measure of abundance lies in the ease and rapidity with which an area can be sampled. Accordingly it has been very widely used as a measure of abundance, often without a realization of the grave disadvantages of the method. The error involved in estimating frequency compared with cover or density is negligible, but the final frequency figure obtained is dependent on several factors, which considerably offset the advantages of speed and facility.

(*a*) THE EFFECT OF QUADRAT SIZE. It is clear from Fig. 1.7 that quadrats *A* and *B* of different size will give very widely differing frequency values

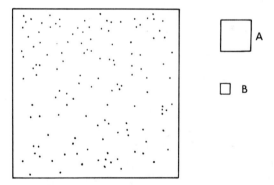

Fig. 1.7. The dependence of percentage frequency on quadrat size. The two sizes of quadrat (*A* and *B*) will give widely differing percentage frequency values from the same diagrammatic community.

for the species represented in the layout. Quadrat A would give a frequency value of about 100 per cent whereas quadrat B would give a frequency value in the low intermediate range. Thus, frequency is dependent on quadrat size and it is important to state the size of quadrat used in an estimate of percentage frequency. Consequently if data are to be compared from different sample plots it is essential to use the same quadrat size throughout.

(b) THE EFFECT OF PLANT SIZE. Again it is obvious that shoot frequency will vary markedly for the two species in Fig. 1.8. Both species have the

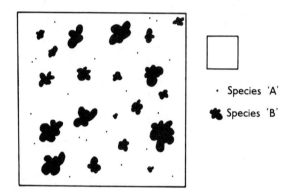

· Species 'A'

🌑 Species 'B'

Fig. 1.8. The dependence of percentage frequency on the size of the individuals sampled. Using the quadrat size given, markedly different frequency values will be found though the same number of individuals of each species are present.

same density but, using the size of quadrat shown, species B will have a high and species A a low percentage frequency. On the other hand if rooted frequency is used both species will have the same frequency value. Again it is essential that the type of frequency measure used is clearly stated.

(c) THE EFFECT OF SPATIAL DISTRIBUTION OF INDIVIDUALS. Three distributions of individuals are shown in Fig. 1.9 all having the same density but a different pattern. With the quadrat size shown the estimated frequency of the species in A will be very high, in B very low and in C intermediate. In A the uniform distribution of individuals makes it almost impossible to take a random sample with zero presence, whereas the marked clumped arrangement in B gives the other extreme.

It is clear from the above considerations that frequency is not only dependent on the density of individuals but also on several other factors whose effect in any given sample area are usually unknown. Thus, despite the desirability of finding a simple conversion factor for transforming frequency data, which are readily obtained in the field, into density data, the situation is so complex as to make it impossible. Under random conditions it is possible to derive a satisfactory relationship (see Greig-Smith 1957), but random distributions of individuals in vegetation are infrequent and such a transformation is of academic interest rather than practical use.

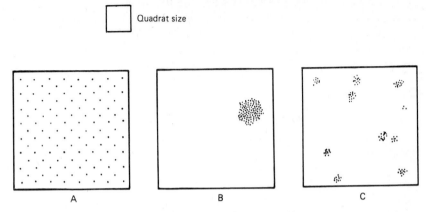

Quadrat size

A   B   C

Fig. 1.9. The dependence of percentage frequency on the pattern of the distribution of individuals. With the same number of individuals present in each case and using the same size of quadrat, widely differing values will be obtained from the three communities.

In general, frequency is a useful measure of abundance where comparisons are to be made on a large scale, as long as the limitations are borne in mind but it would seem worth while (despite the time necessary for the sampling) using some other method wherever possible.

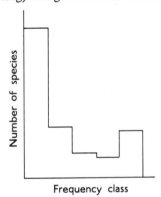

Fig. 1.10. The relationship between frequency class and the number of species falling within each of the classes, showing the typical J-shaped form of the curve.

A further point over which there has been considerable controversy is Raunkiaer's Law of Frequencies often referred to (in its graphical form) as Raunkiaer's *J*-shaped distribution curves (Raunkiaer 1928). The law of frequency simply states that the numbers of species of a community in the 5 percentage frequency classes, i.e. those of 0–20, 21–40, 41–60, 61–80 and 81–100 per cent frequency are distributed as in Fig. 1.10, i.e.

$$A > B > C \gtreqless D < E.$$

The general fall in the first three or four classes is to be expected since there are more rare species in an area than common ones but the increase in the fifth class (E), corresponding to the frequency class 71–100 per cent, is not to be expected and, as Greig-Smith (1957) states, 'It has played a notorious part in some attempts at definition of plant associations'. The increase in class E reflects the theoretical infinite range of density and contrasts with the more strictly defined limits for classes A, B, C and D. This fifth class has a possible density range greatly exceeding that occurring in the remainder of the frequency classes A–D. Accordingly many more species fall into this class, despite the general tendency for 'common' species to be relatively few in number in a community.

Yield

As a measure of abundance, yield has not been used very widely in ecological work. It is determined by clipping a series of sample quadrats, sorting the clippings, drying and weighing them, so that for each species present a figure for dry weight is available. Under field conditions it is very time consuming to sort out clippings into the component species and accordingly its use as a measure of abundance is limited. Where a series of culture experiments or a series of field trials are being undertaken, then yield offers a reliable and absolute measure of both abundance and vigour of a species.

Performance

In some instances it is necessary to know how well a plant is growing in a particular area and a measure of some suitable character can often be made, which reflects the relative vigour or performance, as it is termed, of an individual. Suitable measures include leaf length, leaf width or a ratio of the two, these measures being related to leaf area (the area available for photosynthesis), or flower number, length of flowering spike, number of seeds per capsule, etc., reflecting the reproductive capacity of an individual. The choice of character depends entirely on the species under investigation, but in general leaf length or width is often suitable and gives a good measure of performance. Such a measure was used by Kershaw (1960b, 1962a) on *Alchemilla alpina* and *Carex bigelowii* (pp. 72 to 74). Phillips (1954b) has shown that the ratio of the length of triangulate tip in a leaf of *Eriophorum angustifolium*, to the channelled portion of the leaf, is controlled by the relative growth rates of shoot apex and leaf primordia. As a result this ratio offers a good measure of performance of *Eriophorum*, and its general vigour under different environmental conditions can be readily determined in the field (Phillips 1954a).

2 Sampling,   Tests of Comparison
Application of Quadrat Measures

THE importance of sampling a plant community so as to obtain the maximum amount of information from one set of samples, has been emphasized by numerous workers and Greig-Smith (1957) gives a comprehensive and detailed account of the problems involved. However, it is necessary to outline the more important aspects since a series of conclusions will be based on a single sample, and if the sampling procedure is incorrect, then the derived conclusions may be invalid. The importance of sampling has been stressed to the point where an impression is gained that the procedure is very complicated and in the hands of the uninitiated a dangerous tool. This is only partly true, since the approaches to this problem are all based on common sense. Often the correct procedure and also the one that will give the maximum amount of information on the problem in hand, can be readily devised by careful thought beforehand and with only an elementary knowledge of statistical methods. On the other hand since complex sampling methods are widely used advice from a statistician at an early stage may save wasted labour in the field. The necessity to carefully plan the sampling beforehand is important in yet another sense; recording the contents of a quadrat is nearly always a lengthy and a tedious operation and it is very much better to obtain all the required information from one set of quadrats rather than to have to go back and repeat the sample so as to obtain additional information. Thus no hard and fast rules can be laid down; each sampling procedure should be thought out in relation to a specific problem, the information required being borne in mind.

STATISTICAL METHODS

In all ecological work there is a certain error involved in any field observations or experimental results. This error is partly due to the inadequacies of the equipment, and, more important, to an unknown amount of variation always found in biological material. Thus it is necessary to know whether the difference between, for example, the number of buttercups in two plots is a real difference reflecting some variation in the environment of the two plots, or whether in fact this difference is more than covered by the error attached to the data. In order to assess the reality of the results it is necessary to compare the difference between them with an estimate of the error, or in other words to test the significance of the result.

Normally it is impractical to count every individual present within an area and accordingly a sample of the density is taken from the total population by using a number of randomly placed quadrats.

VARIANCE AND STANDARD ERROR

The following data represents the number of individuals of a species in ten quadrats placed at random in two sample areas:

$$\text{Plot I} \quad 6, \ 3, \ 0, \ 1, \ 8, \ 2, \ 7, \ 0, \ 1, \ 4$$
$$\text{Plot II} \quad 7, \ 11, \ 15, \ 8, \ 14, \ 9, \ 12, \ 14, \ 15, \ 10$$

The average density is obtained by summing the numbers and dividing by the number of quadrats in each plot:

$$\text{Plot I} \quad \bar{x} = \frac{S(x)}{n} = \frac{32}{10} = 3 \cdot 2$$

$$\text{Plot II} \quad \bar{x} = \frac{S(x)}{n} = \frac{115}{10} = 11 \cdot 5$$

The problem with which the observer is now faced is to decide whether the difference between these two mean values has an ecological significance or whether the sampling error is large enough to explain it. Both sets of data show a range of values reflecting the biological variation of the data and the range (8) is almost equal to the difference between the mean values. It would thus be reasonable to account for the difference in the means by this obvious biological variation. The variation of the density readings is expressed statistically as the *variance* of the sample and is calculated as follows:

$$\text{Variance} = \frac{S(x - \bar{x})^2}{n - 1}$$

This is algebraically identical with

$$\text{Variance} = \frac{S(x^2) - \dfrac{(Sx)^2}{n}}{n - 1}$$

The latter offers an easier alternative to the rather laborious subtraction of the individual values from the mean value and squaring the results. The divisor $(n - 1)$ is known as the degrees of freedom of the sample, and is used instead of n to compensate for using the sample mean instead of the true population mean. That is to say, if the whole population had been included in the sample, the mean density would be an accurate value. As it is, only ten quadrats were taken and the mean value thus obtained is only an estimate of the true mean, and accordingly the use of this sample mean will slightly bias the estimate of sample variance. By using $(n-1)$, some compensation for this bias is achieved. Usually the degrees of freedom of a sample are one less than the number of samples taken, though there are exceptions to this. The significance of the difference between the two mean values is assessed by comparing it with the standard error of the difference which itself is related to the variance. The calculation is as follows:

I. $S(x) = 32,\ n = 10,\ \bar{x} = 3\cdot2$

$S(x^2) = 6^2+3^2+\cdots+4^2+1^2 = 180$

$S(x)^2 = 1024$

$$S(x^2)-\frac{(Sx)^2}{n} = 180-102\cdot4 = 77\cdot6$$

$$\text{Variance} = \frac{77\cdot6}{n-1} = 8\cdot6$$

II. $S(x) = 115,\ n = 10,\ \bar{x} = 11\cdot5$

$S(x^2) = 1401$

$S(x)^2 = 13,225$

$$S(x^2)-\frac{(Sx)^2}{n} = 78\cdot5$$

$$\text{Variance} = \frac{78\cdot5}{n-1} = 8\cdot7$$

The estimate of the error of the difference between the two means is obtained from the variance of the total data.

Estimate of variance of the total data is given by

$$V = \frac{\left(Sx_1^2-\frac{(Sx_1)^2}{n_1}\right)+\left(Sx_2^2-\frac{(Sx_2)^2}{n_2}\right)}{n_1+n_2-2} = \frac{77\cdot6+78\cdot5}{18} = 8\cdot67$$

This is the variance of the total data, and the variance of the *difference of the means* accordingly is $V\left(\frac{1}{n_1}+\frac{1}{n_2}\right)$ and the standard error is

$$\sqrt{V\left(\frac{1}{n_1}+\frac{1}{n_2}\right)} = \sqrt{\frac{8\cdot67}{5}} = \sqrt{1\cdot734} = 1\cdot317$$

t-TESTS

The standard error of the difference is now used to assess the difference between the means by a t-test:

$$t = \frac{\text{Difference between the means}}{\text{Standard error of the difference}}$$

i.e. $\dfrac{11 \cdot 5 - 3 \cdot 2}{1 \cdot 317} = 6 \cdot 302$ with 18 degrees of freedom

i.e. the difference between the means is more than six times as great as the standard error of the difference and this difference is highly significant. The level of significance is readily obtained from a table of t (Fisher and Yates 1957). With $n = 18$ the probability of exceeding a value of t of 3·92 by chance is very small ($p = 0 \cdot 001$). Thus with $t = 6 \cdot 3$ the difference between the two mean values arising by chance is very much less than 1000:1 and this difference is highly significant. [A 5 per cent level of significance (1 in 20) is usually taken as the level of statistical significance.]

It is important to realize that a significant value of t does not necessarily mean a real difference between the two values. A large value of t can arise either by the difference between two mean values being significant, or by the variance of the two populations being significantly different, or both. Thus it may be necessary in comparing data to use a modified version of the t-test when the variance of the two samples is significantly different. This is readily confirmed by a variance ratio test. In the above example the two sample variances were 8·6 and 8·7 respectively and the ratio of the larger variance to the smaller is approximately 1. From variance ratio tables entered with 9×9 degrees of freedom this difference in variance is not significant (a variance ratio of 3·2 has to be exceeded before the difference can be regarded as statistically significant) and the use of a t-test is quite legitimate. (For details of the modified t-test, see Greig-Smith 1957, Snedecor 1946.)

CONFIDENCE LIMITS or FIDUCIAL LIMITS

In a normal population with a mean of $\bar{x}$ the shape of the curve simply relates individual sizes or numbers to this mean value. Thus most of the population is roughly equal to the mean value but a proportion of the individuals are much larger or smaller. Just as the variance of the population is a measure of the variability then the *standard deviation* (standard deviation = $\sqrt{\text{Variance}}$) is a linear standard of measurement along the x-axis (Fig. 2.1). The total population of individuals in fact is contained in approximately three standard deviations either side of the mean. The majority of values lie close to the population mean $\bar{x}$ and it can be shown from a table of C (see Fisher and Yates, 1951) that 68·26 per cent of the population falls within one standard deviation either side of the mean. Identical considerations apply to populations of *mean values* where

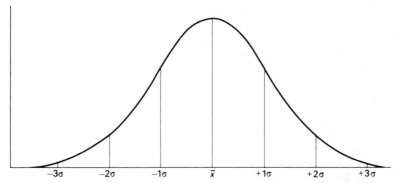

Fig. 2.1. The normal distribution curve with mean $\bar{x}$ and linear scale of unit standard deviations σ.

68·26 per cent of the population of sample means falls within one standard error of the true population mean value (S.E.M.). The 68·26 per cent *confidence limits* are then simply:

$$\text{Upper limit} = \bar{x} + 1 \text{ S.E.M.}$$
$$\text{Lower limit} = \bar{x} - 1 \text{ S.E.M.}$$

Similarly for a 5 per cent level of significance, the 95 per cent confidence limits are:

$$\text{Upper limit} = \bar{x} + 1 \cdot 96 \text{ S.E.M.}$$
$$\text{Lower limit} = \bar{x} - 1 \cdot 96 \text{ S.E.M.}$$

and the 99 per cent confidence limits

$$\bar{x} \pm 2 \cdot 58 \text{ S.E.M.}$$

From the data above

I.   Variance $= 8 \cdot 6$ $n = 10$

$$\text{S.E.M.} = \sqrt{\frac{V}{n}} = \sqrt{\frac{8 \cdot 6}{10}} = 0 \cdot 93$$

and the 95 per cent confidence limits are:

$$\bar{x} \pm 1 \cdot 96 . \text{S.E.M.} = 3 \cdot 2 \pm 1 \cdot 96 \times 0 \cdot 93$$
$$= 3 \cdot 2 \pm 1 \cdot 82$$

The probability is 0·95 that 1·4 and 5·0 include the true population mean.

Similarly the 99 per cent confidence limits are $3\cdot2\pm2\cdot40$. A statement of the confidence limits as a scale mark in relation to graphed data, enables a rapid visual assessment of both the significance of any differences as well as the quality of the data.

χ^2-TEST FOR ASSOCIATION BETWEEN SPECIES, SCATTER DIAGRAMS

Useful information can be obtained from field data on the possible inter-relationships between different species by counting the number of quadrats out of the total in which one or both of a pair of species are found. For example in n quadrats, a contained species A and B, b species B alone, c species A alone and d quadrats contained neither species A or B. A 2×2 contingency table is then made up as follows:

SPECIES A

		+	−	
SPECIES B	+	a	b	$a+b$
	−	c	d	$c+d$
		$a+c$	$b+d$	n

It can be calculated that the number of quadrats containing both A and B, provided they are not associated in any way, should be $[(a+b)(a+c)]/n$. Thus we have a the observed number of quadrats containing both species and we can calculate the expected number assuming an independent relationship between the two species. The other expected frequencies of the different A and B quadrat combinations can readily be obtained simply by subtraction from the marginal totals. Thus the expected number of quadrats $+A$, $-B$ would be

$$(a+c)-\frac{(a+b)(a+c)}{n} \quad \text{and so on.}$$

The number of observed quadrat combinations of A and B can now be compared with the expected number by a χ^2-test.

The calculation of χ^2 is conveniently done from the formula

$$\chi^2 = \frac{(ad-bc)^2.n}{(a+b)(c+d)(a+c)(b+d)}$$

the χ^2 tables are entered with one degree of freedom. The χ^2 value shows whether the difference between the observed and expected numbers of quadrats is significant or merely due to a chance fluctuation. The trend of the association, i.e. whether the association is positive or negative, is given by a comparison of the expected and observed values for the joint occurrence of A and B. Thus if A and B occur together *more* frequently than the expected number of joint occurrences, then A and B are positively associated. Similarly if A and B occur together *less* than expected they are negatively associated.

The level of association between species is very quickly calculated but since the data is basically a measure of frequency the result obtained is dependent on quadrat size. Consider the relationship of species A, B and C as shown below (Fig. 2.2). Clearly by inspection A and B are markedly positively associated; A and C and similarly B and C, negatively associated. Now consider the effect of sampling in a random fashion such a population with quadrat size I, roughly equal in size to the individuals A, B and C. Most quadrats would contain either A or B or C or none of them, but hardly any quadrats with AB, AC or BC. Accordingly the χ^2-test would

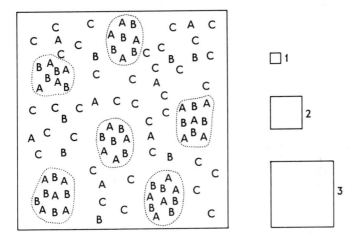

Fig. 2.2. The relationship between quadrat size and trend of association between three species A, B and C (see text).

show strong negative association between all species. With quadrat size 2 A and B would on the average be found frequently together and species C on its own. A χ^2-test would confirm the obvious trends of association between A, B and C. However consider finally the sampling of the layout with quadrat size 3. Nearly all quadrats would contain species A, B and C and, in this instance, the number of quadrats containing neither of the pairs of species would be zero and χ^2 cannot then be calculated. However, assume the same pattern but with a much lower density of individuals, then the final size of quadrat would show nil association or possibly a positive association between all paired combinations of the three species. It is possible to get three trends of association merely by changing the size of sample quadrat and considerable care must be taken in interpreting a set of positive and negative associations.

Obviously in this instance the pattern of distribution relative to the size of quadrat is important and, since most if not all vegetation contains pattern (see Chapter 8), the trend of association will always vary considerably with different sizes of quadrats. As Greig-Smith (1957) points out, very little attention has previously been directed to this important point and yet it has far reaching effects in the considerations of statistical analysis of community groupings (p. 168). If however the scale or scales of pattern are known, then the variation of trend of association between two species with increasing quadrat size gives valuable information as to the detailed inter-relationships of the species in the area examined.

When data is in the form of a quantitative measure as opposed to mere presence and absence, then a correlation coefficient is a more appropriate measure of the degree of association; details of the calculation are obtainable in any book of elementary statistics (see also Chapter 5).

An indication of a possible relationship between two factors can be very easily obtained by means of a scatter diagram. An example is given in Fig. 2.3 showing the marked relationship between pH and percentage base saturation. The relationship is so clear-cut that it is probably unnecessary for a correlation coefficient to be calculated; where the scatter of points becomes greater and the degree of correlation correspondingly lower, the visual confirmation of a suspected correlation becomes impossible and it is necessary to calculate the actual coefficient. A scatter diagram showing the relationship between percentage cover of *Festuca* and *Agrostis* in chalk grassland (see also Fig. 2.7) is given in Fig. 2.4, and illustrates a more typical degree of scatter. (The actual data is given in Table 5.1, p. 86.) In most cases it is worthwhile to calculate the correlation coefficient since it can be very tempting to place too much weight on an apparent correlation but one that only has a few degrees of freedom.

SAMPLING METHODS

RANDOM SAMPLING

It is usually necessary to distribute a set of quadrats in an area in such a way that the position of each quadrat is independent of all the other quadrats and is also independent of any prominent feature of the area. This may be made clear by considering the hypothetical case where a series of quadrats are being taken from a sample plot containing a few large and conspicuous thistles. Subconsciously, if not consciously, the very size of the thistles will lead to a feeling that they *really ought* to be included

Fig. 2.3. The relationship between pH and base saturation expressed as a scatter diagram. (From Gorham 1953; courtesy of *J. Ecol.*)

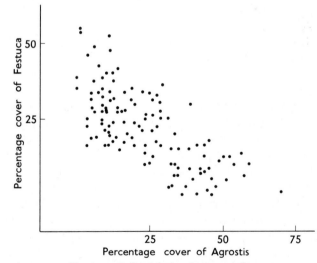

Fig. 2.4. A scatter diagram of the relation between the percentage covers of *Festuca* and *Agrostis* in chalk grassland.

in the sample. If the randomization of the quadrats is being attempted by the closing of eyes or throwing over the left shoulder, then inevitably the tendency is to try to include at least *one* thistle in a quadrat throw. This leads to over-sampling of the area closely adjacent to the thistles. This error is surprisingly common and very often leads to completely biased sampling since the effect of *trying* to make a random throw often produces a final distribution of quadrats which are too regularly distributed. Thus conscious or unconscious prejudgement has to be eliminated, otherwise the final non-random sample will be of little use for further comparison. The use of a *t*-test, or the calculation of any standard error, demands that the original sample should be taken at random. If there has been any introduction of non-random sampling then the calculated standard error is a fictitious figure and a statistical comparison is not valid.

The most satisfactory way of obtaining a random sample from an area is to lay out two lines at right angles to each other to serve as axes; each line is then marked by a convenient number of divisions. A series of random numbers taken from tables (or merely by drawing a set of numbers each time from a pack of cards) then are used as pairs of co-ordinates to locate each quadrat in turn (Fig. 7.1, p. 130). It is often quicker to mark all the random points by a series of pegs before enumerating the quadrats, each peg marking a corner of the quadrat.

Size of sample necessary

It is impossible to make a general rule as to the number of sample quadrats or measurements to take from an area. Each case has to be decided independently and considered against the amount of sampling involved to reach a certain level of accuracy. It is possible to obtain a subjective assessment at least of the size of sample necessary for any given problem. The method consists of taking a sample of five quadrats or readings and calculating a mean value. The number of samples is then increased to 10, 15, 20 . . . and a fresh mean is calculated each time. A graph relating mean value to the sample size is constructed and the point (i.e. the sample size) at which the mean value ceases to fluctuate is easily determined (Greig-Smith 1957). An example is given in Fig. 2.5 for performance data of *Calamagrostis neglecta* expressed as leaf length. The curve fluctuates considerably up to a sample size of about 50 after which there is little variation and a sample size of 80 would give a reasonably accurate measurement of mean performance in that particular site. (For additional methods of estimating the required sizes of sample see Greig-Smith 1957, Snedecor 1946, etc.)

It is obvious that as the sample size is increased a better measure of the mean of the population is obtained, but to reduce the standard error the

number of samples has to be increased greatly and in general it is recommended to take as large a sample as time will permit. The calculation of a
series of mean values as the sample size is increased merely acts as a guide
and even if the fluctuations are reduced markedly at a given sample size it
does not mean that the fluctuations have ceased altogether. In the case

Fig. 2.5. The reduction of the variations in the value of the mean, as the size of
sample is increased (see text).

quoted above additional samples would continue to show a slight fluctuation but certainly less than 0·1 mm and since the measurements were
taken to the nearest millimetre it would have been pointless to increase the
sample beyond 100. Similar considerations can be applied to quadrat
density, frequency or percentage cover. However, if very slight differences
are suspected to be of ecological significance then a large sample will be
unavoidable.

Size and shape of quadrat

  If the distribution of individuals in a population is completely random
then the size of quadrat used is immaterial except from the standpoint of
convenience. Any size convenient for the particular vegetation which is investigated can be used—a small quadrat if the species are small and
numerous so that the counting of individuals is made easier, or a large
quadrat if the species are large or thinly scattered. However most individuals are not distributed at random and the size of the quadrat has a
considerable effect on the variance of the data obtained. This is discussed
more fully later (p. 136) but, briefly, if there is a tendency for individuals

to be grouped together, the measure of the variance of the data is at a maximum when the size of quadrat is approximately equal to the mean area of the groups of individuals. Since it is usually impossible to pre-determine the scale of this pattern (as it is termed) and since such patterns are often repeated on larger scales, the size of quadrat has to be chosen quite arbitrarily. The most suitable quadrat on theoretical grounds being the smallest possible, relative to the type of vegetation and to the practicability of the enumeration of such a quadrat size.

The shape of a quadrat has been a square almost by tradition but some advantages are obtained by the use of rectangular quadrats. Clapham (1932) showed that the variance between rectangular strips was less than between squares of the same area, in one instance at least. Presumably this resulted from a rectangle crossing a high density patch and an adjacent low density patch simultaneously and thus levelling out the variance over the whole area. However, this would not always be true and the only con-sistent advantage for the use of rectangular quadrats is the increased facility with which the quadrat can be studied. Thus in large quadrats there is a tendency to crush part of the vegetation by trampling or leaning and this can be a serious disadvantage when several types of measurements are being taken from the same quadrat. One common error associated with quadrat work which is often not appreciated is the 'edge effect'. Fre-quently a decision has to be taken as to whether a species on the edge is 'in' or 'out' of a quadrat. Where the edge of a quadrat is large relative to the area inside it, this error can be quite marked and thus very long thin quadrats, or conversely very small square quadrats should not be used.

REGULAR SPACED SAMPLES

From the considerations outlined above it is quite clear that in the majority of cases samples of a population taken at random with their re-sultant standard error are of considerably more use than samples taken at regular intervals. However, there are some exceptions to this general rule. Where data about the abundance of a species along an environmental gradient are required, it is worthwhile to sample systematically along a transect (p. 33) in order to record the abundance of a species in relation to position. Similar considerations apply to isonomes (p. 39) and to the grid analysis of pattern (Chapter 7, p. 128) both of which relate the abundance of a species to location of the quadrats, the former more as a descriptive method, the latter as a statistical analysis of non-randomness. On the other hand regular sampling can lead to a considerable bias of the mean value as would occur in the case in Fig. 2.6. This is a rather extreme example of the hypothetical distribution of a species on furrows and ridges which developed in relation to the drainage of a field. The regularly spaced sample points will coincide exactly with the equally regular furrow and ridge, and the mean value for the abundance will be considerably different from that obtained by random sampling.

A similar false impression can be obtained for the optimum level of pH of a species if the sample is not randomized. Thus numerous studies have

←——— *Sample interval* ———→

Fig. 2.6. A hypothetical ridge and furrow distribution of a species which when sampled at regular intervals would produce a very biased estimate of abundance.

been made relating the distribution of a given species to pH by taking a soil sample at each site examined where the species was growing. This in fact is a sample of the variation of soil pH in a given area and the resultant data relates to the frequency distribution of pH in that particular area. Such data offers no real proof of an interrelationship between a species and pH since samples are not randomized, although in fact such a relationship may exist. Emmett and Ashby (1934) present a set of data for *Pteridium aquilinum* and *Vaccinium myrtilis* (Table 2.1) obtained from a

Table 2.1 Frequency distribution of occurrence of *Pteridium* and *Vaccinium* in classes of pH. (From Emmett and Ashby 1934; courtesy of *Ann. Bot., Lond.*)

pH class means	Number of samples	Frequency of *Pteridium*	Frequency of *Vaccinium*	Percentage frequency of *Pteridium*	Percentage frequency of *Vaccinium*
4·8	2	0	2	0	100·00
4·9	2	0	2	0	100·00
5·0	2	0	2	0	100·00
5·1	5	1	4	20·00	80·00
5·2	13	6	8	46·15	61·54
5·3	17	8	11	47·07	64·71
5·4	7	4	3	57·14	42·86
5·5	44	27	28	61·36	63·64
5·6	78	41	54	52·56	69·23
5·7	16	5	12	31·25	75·00
5·8	7	5	1	71·43	14·29
5·9	9	4	3	44·44	33·33
6·0	7	5	1	71·43	14·29
6·1	3	2	0	66·67	0

random sample. The most frequently occurring pH value in the samples is pH 5·6 and it is not surprising that the frequency of occurrence of *Pteridium* and *Vaccinium* are apparently correlated with this particular soil pH as an optimum for both species. In fact the data merely demonstrates three frequency distributions present in an area which may or, equally well, may not be correlated. The two final columns of the table give the percentage occurrence of the two species, and are derived from the second, third and fourth columns, for example at pH 5·1 one sample out of five contained *Pteridium* which is in fact a 20 per cent occurrence.

The data presented in this form show immediately the dissimilar relationship of each species with pH. It is probably worthy of note that the most efficient way of approaching this particular problem is to take a series of random samples noting the pH and abundance of the species in question. A correlation coefficient can then be readily calculated.

PARTIAL RANDOM SAMPLING

If an area on inspection appears to be heterogeneous, it is pointless to sample at random and to obtain what may be merely an arbitrary set of figures of species abundance. In such a circumstance the recognized method is to sub-divide the area into a convenient number of equal sized areas and to take random samples within each sub-division. If the area proves to be uniform after all then the data can all be lumped together, but on the other hand if the area is variable then information is available which is relative to each sub-division of the plot and is accordingly of considerable interest. The use of partial random samples would be equally applicable to the hypothetical relationship outlined above between the distribution of a species and topography (Fig. 2.6) and the approach in general is ideally suited for sample plots which contain a slight variation of surface topography.

TRANSECT AND ISONOME STUDIES

Transects are of considerable importance in the description of vegetative change along an environmental gradient, or in relation to some marked feature of topography. The method simply consists of laying out a line running across the zones to be sampled and then placing quadrats at known intervals along the line. Various measures can be employed in the description of a transect, density or frequency determinations being made at intervals along the line, and cover determination most conveniently made by using a frame of pins orientated at right angles to the transect at fixed intervals. The data is suitably represented, either by means of a graph or a histogram, of abundance of a species plotted against position on the transect.

In Fig. 2.7 the transect data illustrating the change of composition of grassland on the North Downs is presented; the transect runs from the zone of acidic clay with flints down into neutral or slightly alkaline chalk grassland. The data is in the form of percentage cover, 80 points being taken from each 80 cm strip of the transect and expressed as a percentage. The general distribution of the species in relation to each other and to the slope of the area can be readily appreciated.

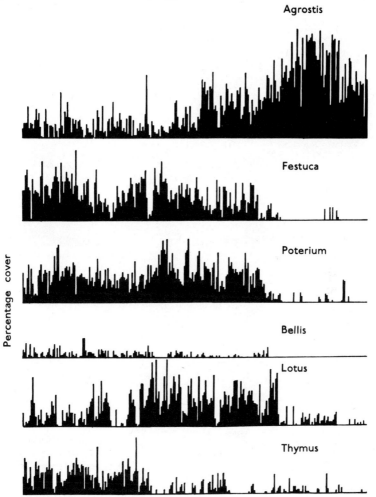

Fig. 2.7. A series of histograms of percentage cover for six species of chalk grassland showing the relationship between the abundance of the species along a transect.

A further use of transect studies is in the description of zonation representing stages of a succession and Alvin (1960) gives data on the frequency of different lichen species occurring in different zones of sand-dune vegetation (Fig. 2.8). The sampling in this case was done by dividing the dune system into six zones, 100 quadrats being enumerated in each of the zones. Again the distribution of lichen species is readily appreciated and the general succession from *Cladonia coniocrea, C. cornutoradiata*, etc., on the first ridge to *C. coccifera, C. squamosa, C. crispata*, etc., on the third ridge is quite clearly demonstrated.

Fig. 2.8. Histograms representing the distribution of the main lichen species in six zones of sand-dune system. One division of the vertical scale represents 25 per cent frequency. (From Alvin 1960; courtesy of *J. Ecol.*)

In general the use of a transect is of greater importance where the variation of vegetation in response to a changing environmental factor is fairly well marked. If the change is only slight it is then necessary to lay out a number of sample plots at intervals along a transect and make a detailed analysis by means of random quadrats, within each plot. By this means slight trends in the density of a species down a slope may be detected and if necessary correlated with environmental factors which are measured at the same time.

An analogous approach to variation in the spatial distribution of a species in a sample plot can be made by the use of the isonome method outlined by Ashby and Pidgeon (1942). The method consists of laying out a grid of contiguous quadrats over a sample area and recording for each quadrat the abundance of the species to be considered. For each species separately the density distribution is then reproduced on squared paper and those squares with approximately equal abundance values joined by a series of lines, giving a 'contoured' distribution of species abundance. This method can be extended to suitable measurements of the environment so that the distribution of a species in an area can be related to the distribution of other species or to environmental factors. The method may be made clearer by considering an actual example.

The microtopography of an area of hummocks in limestone grassland at Malham Tarn, Yorkshire, is presented in the form of an isonome (Fig. 2.9) constructed from a series of height measurements taken from a levelled datum line. At the same time the density of *Carex flacca* and *C. panicea* were recorded for each unit of the grid and the corresponding isonomes constructed from the density data for both species. The marked relationship of both *Carex* species with the hollows, is readily appreciated and the isonome method is in general a very convenient way of depicting relationships between species or between a species and some environmental factor. Clearly the method is only applicable where the distribution of species varies quite markedly; where the distribution is not readily appreciated by the visual inspection of an isonome it is necessary to use a more elaborate method (see Chapter 7).

Both transect and isonome methods illustrate the use of regularly spaced samples in describing an area and it should be again emphasized that the sampling procedure most suitable for a given problem is usually chosen or designed for that individual problem. The choice of measure, the size of quadrat, the size of sample and the random or regular position of the quadrats can all be decided more by common sense than resorting to complex statistical theory.

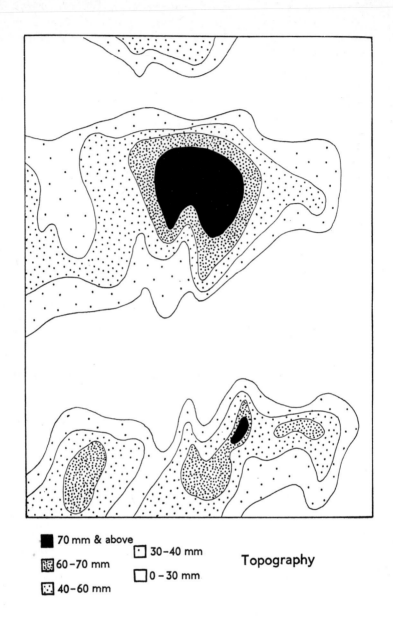

70 mm & above

60 -70 mm

40 - 60 mm

30 - 40 mm

0 - 30 mm

Topography

DENSITY OF CAREX FLACCA

■ Density > 2   ⊡ Density 2   □ Density 0 - 1

Fig. 2.9a. Isonomes of the density of Carex flacca showing the marked relationship between their relative abundance and micro-topography.

■ Density > 2   ▦ Density 2   □ Density 0 – 1

DENSITY OF CAREX FLACCA

Fig. 2.9a. Isonomes of the density of *Carex flacca* showing the marked relationship between their relative abundance and micro-topography.

■ Density > 2　　▒ Density 2　　▢ Density 0 – 1

DENSITY OF CAREX PANICEA

Fig 2.9b. Isonomes of the density of *Carex panicea* showing the marked relationship between their relative abundance and micro-topography.

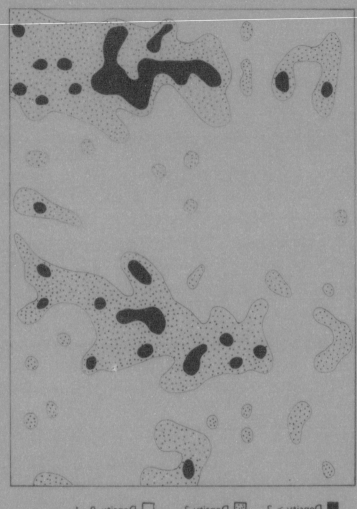

DENSITY OF CAREX PANICEA

■ Density > 2   ▨ Density 2   □ Density 0 – 1

Fig 2.9b. Isonomes of the density of Carex panicea showing the marked relationship between their relative abundance and micro-topography.

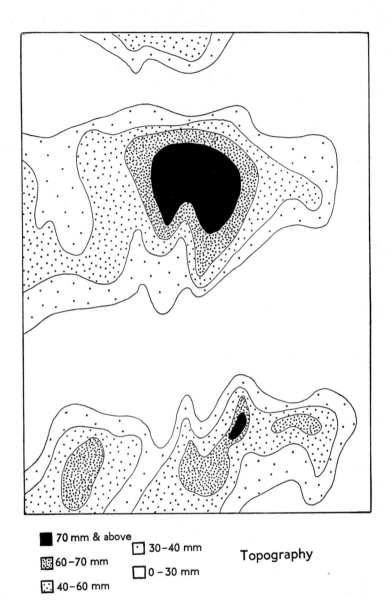

70 mm & above

60–70 mm

40–60 mm

30–40 mm

0 – 30 mm

Topography

3 Vegetational Change. Plant Succession and the Climax

SUCCESSION

AN area of bare ground when stripped of its original vegetation by fire, flood or by the drainage of a lake, does not remain devoid of vegetation for very long. The area is rapidly colonized by a variety of species which will subsequently modify one or more environmental factors. This modification of the environment in turn allows further species to become established. A subsequent development of the vegetation, by this reaction of the vegetation on the environment followed by the appearance of fresh species, is termed *succession*. The concept of succession was largely developed by Warming (1896), and Cowles (1901) who related the stages of sand-dune development in a time series, and finally by Clements (1904) who outlined in considerable detail the stages of a large number of plant successions, initiated by a variety of factors (Clements 1916).

Clements introduced the term *sere* to describe the developmental stages through which the vegetation passes until it reaches an ultimate state of equilibrium with the climate and major geological factors of the area. Thus the well known colonization of open water is termed the hydrosere and includes several marked *stages*, starting with submerged aquatic plants and followed successively by floating-leaved aquatic plants, reed swamp, marsh or fen and finally carr. Clements similarly termed the stages of salt marsh succession a halosere, sand-dune succession a psammosere and the development of vegetation on bare rock surfaces a lithosere.

Analysis resolves the process of succession into several important phases: 1. Nudation, which is the initiation of the succession by a major disturbance in the environment. 2. Migration of available species (migrules) to fill the vacant ecological niches. 3. Ecesis or the subsequent ability of the migrules to germinate, grow and reproduce successfully. 4. Competition. 5. Reaction. 6. Final stabilization. (Clements 1916.)

The reaction of species on the environment is one of the outstanding features of succession and constitutes the major mechanism of environmental change which allows further migrules to enter. A very obvious example of such a reaction is found in the formation of sand-dunes. The

establishment of a rhizome fragment of *Ammophila arenaria* and the subsequent production of aerial shoots constitutes a disturbance of the environment at that point. The aerial parts offer an increased impedance to
the wind, any sand particles carried by the wind being deposited round the
plant and gradually burying it. As the sand accumulates, continued growth
of the *Ammophila* keeps it at a generally higher level than the low mound
that is formed, and accumulation of wind-borne sand continues. Production of axillary buds and the development of the typical clumped habit of
Ammophila accelerates the rate of 'dune' formation and gradually over a
period of years the dune gets progressively larger (Fig. 3.1). This is a very

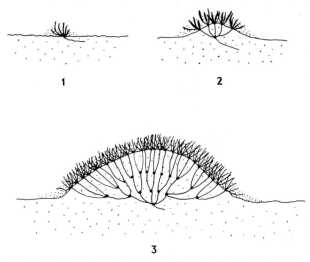

Fig. 3.1. Dune formation by the gradual deposition of wind-carried sand
particles around the aerial shoots of *Ammophila arenaria*.

marked example of reaction of a plant on the environment and is comparable to the increase in the rate of silting of a lake by the early colonisers,
which constitute the first stage of a hydrosere. The same mechanism
operates, that is increased impedance to water-borne particles. These are
deposited around the vegetative parts of the plants. The depth of water decreases over a period of years until other species with their vegetative parts
above water begin to invade. The rate of silt accumulation (plus vegetable
remains from the plants themselves) increases and opens the way to the
gradual development of reed swamp and then fen vegetation.

These two examples of reaction are fairly obvious, but frequently the
reactions which are contributing to the impetus of what is clearly succession are far from clear. Vegetational change was divided by Clements into
two kinds: *Primary succession* initiated on a bare area where reactions are,

initially at least, often fairly clear, and *Secondary succession* initiated by a major environmental disturbance and disrupting a previously initiated succession or producing a marked modification in stable vegetation. For example removal of grazing from grassland, or a forest fire, both initiate a secondary succession and it is in such examples of succession that re-actions are often very obscure. It may be that ecesis and competition play a more decisive role in such instances.

Other classification systems have sub-divided secondary successions into a varying number of divisions, but in general little is gained by a complex classification of successions.

The direction in which a succession will develop has been argued at considerable length in ecological literature. Phillips (1934) states 'Succession is *progressive* only; all examples of apparent retrogression are explicable in terms of some disturbing agency'. Similarly *true* succession is defined by Michelmore (1934) as inherent succession of the vegetation which takes place through the plants altering their environment. This is to be distinguished from physiographic succession due to an initial change in the environment which should apparently not be included. Similarly Tansley (1916, 1920) employs the terms autogenic and allogenic to distinguish between these two distinct successional causes. For an understanding of succession at a level which serves as a *useful* concept, such a degree of fine definition is not essential. Any *directional change* in vegetation, be it due to intrinsic properties of the plants, or to changing extrinsic factors (the environment), gives a useful and workable definition of succession as a dynamic process. This was admirably expressed by Cooper (1926) who stated that 'Succession is the universal process of change which is embodied in the great vegetational stream; all vegetational changes, whether internally or externally induced, whether gradual or abrupt, are therefore successional.' (See also Gleason 1926.)

Numerous examples of succession have been described in detail, but very few cases are available where the succession is related to even an approximate time scale. One or two outstanding exceptions to this have appeared in the literature, notably colonization of glacial moraines in Alaska where the age of the moraines is known with a fair degree of exactness. This enables the plant succession and attendant modification of the substratum by the vegetation to be related to a time scale. Similarly the sand-dune succession in the Michigan Lake Basin in America has been studied in detail and related to a time scale obtained by radio-carbon dating methods.

PLANT SUCCESSION ON GLACIAL MORAINES IN ALASKA

The Glacier Bay region in south-east Alaska is typical of the large-scale glacial retreat that has occurred over the last 200 years in North America, Europe and Scandinavia. The position of the snout of the glacier in historic times can be estimated by growth ring counts of the present spruce forest. Thus ring counts of trees growing beyond the maximum extent of glacier activity (easily recognized from aerial photographs) show the oldest trees to be over 600 years of age and growing on a 'raft' of fallen submerged trunks. The oldest trees on the last morainic ridge are approximately 200 years old, whilst the maximum age of trees on moraines nearer the present site of the glacier progressively decreases. Cooper (1937), Field (1947) and Lawrence (1953) all contributed to a detailed study of the recession of the glacier and a clear picture is now available (Fig. 3.2).

Fig. 3.2. The Glacier Bay fiord complex showing the rate of ice recession. (From Crocker and Major 1955; courtesy of *J. Ecol.*)

Their findings are confirmed by comparison of written records extending back to the 1890's and photographic surveys commenced around 1916.

The stages of succession have been described by Cooper (1923, 1931, 1939) and by Lawrence (1953). The pioneer stage is characterized by *Rhacomitrium canescens* Brid., *R. lanuginosum* (Hedw.) Brid., *Epilobium latifolium* L., *Equisetum variegatum* Schleich., *Dryas drummondii* Rich. and *Salix arctica* Pall. The next stage is the appearance of *Salix barclayi* And., *S. sitchensis* Sans. and *S. alexensis* (And.) Colville, which start as prostrate forms eventually developing an erect habit and forming dense scrub. *Alnus crispa* Ait. Pursh. dominates the next stage of the succession and eventually forms almost pure thickets (with some scattered individuals of *Populus trichocarea* T. and G.), which are finally invaded by *Picea sitchensis* Carr. *Picea* initially forms pure stands, but over a period of time

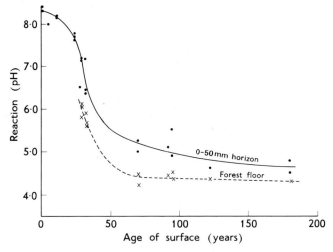

Fig. 3.3. pH of litter residues and surface horizons of the mineral soil with increasing age of surface. (From Crocker and Major 1955; courtesy of *J. Ecol.*)

Tsuga heterophylla Sarg. and *T. mertensiana* Carr. enter the community. On well drained slopes this stage represents the climax vegetation of the area. However, where the ground is gently sloping or flat *Sphagnum* species subsequently invade the moss mat of the forest floor and through the growth of *Sphagnum* and its ability to retain large amounts of water, the forest floor becomes soggy and root aeration decreases. This leads to the elimination of the tree species and to the establishment of muskeg dominated by *Sphagnum* species, with occasional scattered individuals of *Pinus contorta* Laws, which can apparently tolerate the wet conditions which prevail.

Crocker and Major (1955) have studied the time sequence of this succession and the reaction of each stage on the properties of the soil. In the earliest stages of soil development following glacial recession the processes involved appear to be greatly affected by the type of vegetational cover.

The reaction of the soil parent material is quite high with a pH of 8·0–8·4, due to the occurrence of marble in the geological formations of the area. The soil pH falls rapidly with the establishment of the vegetation (Fig. 3.3) and furthermore the rate of increase of acidity is markedly affected by the type of vegetational cover. Bare surfaces show almost no change of pH over a 30-year period, while *Populus, Dryas* and *Salix* sites show a more marked decrease of pH, and *Alnus* produces a striking acidifying effect, lowering the pH of the surface soil from about 8·0 to 5·0 in 30–50 years (Fig. 3.4). This rapid decrease of pH ceases with the invasion

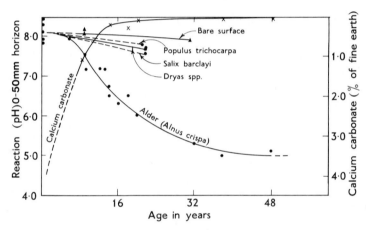

Fig. 3.4. Rate of change of pH in 0–50 mm horizon relative to different vegetational cover and rate of calcium carbonate change under *Alnus*. (From Crocker and Major 1955; courtesy of *J. Ecol.*)

of *Picea*, the leaf litter of this species appears to be approximately of equal acidity to *Alnus* and has no significant effect on the mineral soil.

The organic carbon and total nitrogen concentration in the soil shows equally marked change with time (Figs. 3.5 and 3.6). The vast increase of soil nitrogen is related to the presence of bacterial nodules on the roots of *Alnus* which actively fix atmospheric nitrogen and contribute largely to the reserves of nitrogen in the soil via leaf fall (Fig. 3.7) (Lawrence 1958). The invasion of spruce at this stage and the subsequent reduction of soil nitrogen strongly suggests this to be the important consideration controlling the time of entry of the spruce into the alder thickets. The

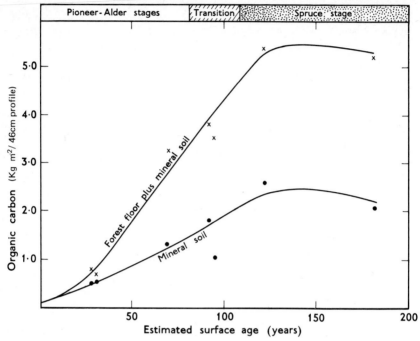

Fig. 3.5. Organic carbon accumulation in mineral soil and forest floor. (From Crocker and Major 1955; courtesy of *J. Ecol.*)

elimination of the alder and its root nodules decrease the net annual addition of nitrogen as leaf litter, for spruce has no such root structures and accordingly the soil nitrogen is reduced.

The complete succession from bare morainic detritus to mature spruce forest takes approximately 250 years during which time the most significant reaction of the vegetation appears to be the building up of soil nitrogen. The reduction of soil pH is probably also closely related to this. The building up of organic carbon will influence to a considerable extent the development of soil structure and will lead to a soil with 'crumb-structure' which is markedly different from the rather amorphous glacial till. In turn this will affect soil aeration and movement of water in the soil, and both of these factors probably influence the succession. However, very little data is available which would enable a general assessment of the effect of such factors on vegetation apart from the rather more obvious relationships that exist. Clearly a detailed study is required of the exact environmental requirements of each species contributing to the succession before the complete mechanism of the succession can be fully appreciated.

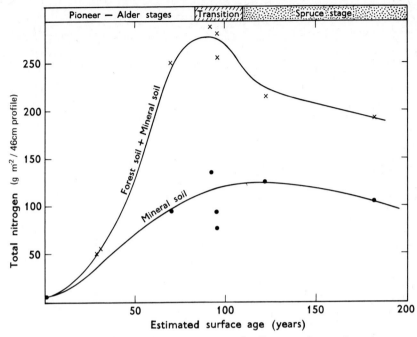

Fig. 3.6. Change of total nitrogen content of soils on surfaces of varying age. (From Crocker and Major 1955; courtesy of *J. Ecol.*)

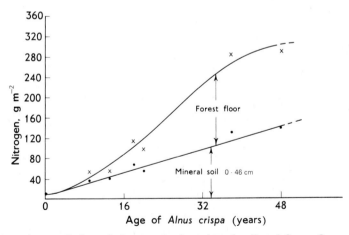

Fig. 3.7. Accumulation of nitrogen in the mineral soil and forest floor under *Alnus crispa*. (From Crocker and Major 1955; courtesy of *J. Ecol.*)

LAKE MICHIGAN SAND-DUNE SUCCESSION

Cowles' (1899) classic description of the sand-dune vegetation of Lake Michigan was one of the early contributions to the concept of plant succession. More recently Olson (1958) has re-examined the successional stages in this area in relation to both the development of the soil and to the age of each dune system.

During and after the retreat of the ice-age glaciers from the Great Lakes region the resultant fall in the general level of the lakes left several distinct 'raised beaches', with their associated dune systems. These systems, running parallel to the present shore line of Lake Michigan are about 8, 13 and 18–20 m above the present level of the lake. These old dune systems have all been dated recently by radio-carbon methods and the highest system above the present lake level (Glenwood) turns out to be slightly over 12 000 years of age. The more recently established dunes were aged by morphological studies of the *Ammophila* colonizing them and growth ring counts of *Pinus banksiana* which enters the succession at an early stage (Fig. 3.8).

The pioneer stage of the succession is dominated by *Ammophila breviligulata* which occasionally becomes established from seed but normally invades freshly deposited sand by rhizomes from young previously established dunes. The rate of dune formation established from a morphological examination of *Ammophila* is rapid and stable dunes are formed in about 6 years. Dunes which have been established for 20 years or more, still have *Ammophila* but with an obviously reduced level of vigour. This loss of vigour has been related to potential leaching of nutrients over a 20-year period. Another suggestion invokes the inherent tendency of this species for internodal elongation which does not cease completely when stabilization has been reached. This results in the growing apex being pushed into the dry surface layers of the sand which are unfavourable for plant growth, with the consequent reduction of vigour of the plant as a whole. A further possibility may be found in the average age of the plants themselves, the loss of vigour being due to the slowing down of physiological mechanisms, resulting directly from old age (see Chapter 4, p. 65). The next stage of the main line of succession is the arrival of *Pinus banksiana* and *P. strobus* which enter immediately dune stabilization is reached. These species eventually are replaced by the final stage dominated by *Quercus velutina*. This main succession is summarized in Fig. 3.8 together with the modifications present in those dune systems where local variation of the environment can change the course of the succession markedly. Thus damp depressions caused by impeded drainage develop into a grassland community and, on the other hand, sheltered pockets on the lee slopes progress to a *Tilia americana* dominated intermediate stage.

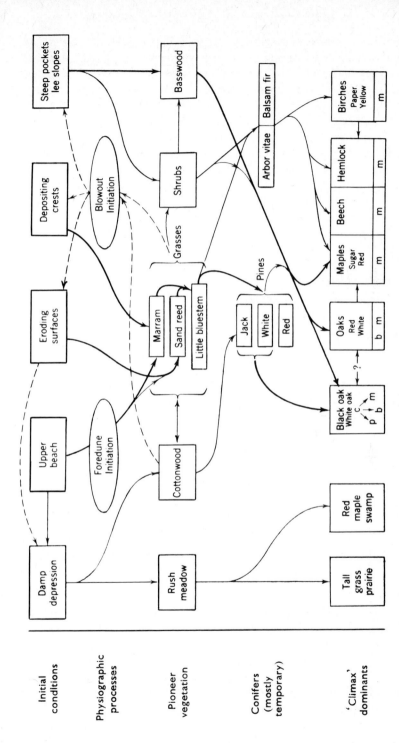

Fig. 3.8. Schematic alternative dune successions in the Lake Michigan dune systems. (From Olson 1958; courtesy of *Bot. Gaz.*)

This different course of succession is related, in part at least, to the micro-climate of these pockets which are largely protected from excessive solar radiation and wind effects. Accordingly moisture availability will be higher, which in addition to the obvious relation to plant growth would contribute some protection from fires which occur frequently in the area, the relatively moist pockets tending to be by-passed. In addition these dune pockets tend to accumulate fallen leaves brought in from the more exposed parts of the dune system and the general nutrient status of the soil increases at the expense of the rest of the area. The reaction of the vegetation on the properties of the soil is broadly similar to the example discussed above but the rate of change is considerably slower. The leaching of carbonate from the soil is very rapid (Fig. 3.9) and the surface layers (10 cm) are more or less devoid of carbonate from the soil after a few hundred years. There is a slow building up of organic carbon (Fig. 3.10) and soil nitrogen (Fig. 3.11). These three environmental changes are virtually completed in 1000 years and subsequent generations of *Quercus* have little effect on these soil properties.

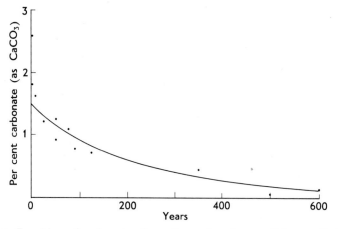

Fig. 3.9. Leaching of carbonates from the surface layers of dune soil. (From Olson 1958; courtesy of *Bot. Gaz.*)

Again the relationships between the environment, the vegetational stages and the development of the succession through modification of the environment can only be described in a general way. The initial dune formation is clear and results in an ecological niche with low fertility which is in fact invaded by species with low nutritional requirements. The subsequent accumulation of soil nitrogen and organic carbon presumably control to some extent the time of entry of oak into the succession. What evidence

Fig. 3.10. Selected profiles of organic carbon from four dunes of contrasting age. (From Olson 1958; courtesy of *Bot. Gaz.*)

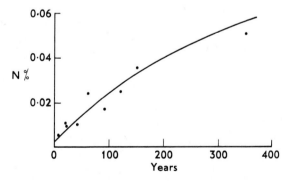

Fig. 3.11. Increase of soil nitrogen in surface layers of soil from dunes of different age. (From Olson 1958; courtesy of *Bot. Gaz.*)

there is suggests that *Quercus velutina* has relatively low nutrient requirements and accordingly able to thrive in a habitat which remains at a very
low level of soil fertility. The annual addition of leaf litter from the final
stages of the succession appears to be equivalent to the amount of nutrient
required to maintain the *Quercus* communities. Thus the soil conditions
remain fairly constant over a long period of time (probably an indefinite
period) and limit the further development of the succession. It appears
very improbable that species with higher nutrient requirements will ever
enter the succession and the occasional and isolated hickory, beech and
maple trees are restricted to sites with special topographic or drainage
features. Accordingly the *Quercus velutina* represents the final stable
vegetation of the area largely restricted by the edaphic features outlined
above.

THE CLIMAX STATE

In both the examples of succession described above the vegetation has
developed to a certain stage at which point it has reached a state of equilibrium with the environment. This final stage of the succession is related
directly to the environment and is referred to as the climax. Numerous
definitions of the climax have been made: Braun-Blanquet (1932) defined
it as the development of vegetation and the formation of soil towards a
definite end point determined and limited by climate. Similarly Odum
(1953) suggests that a climax is the final or stable community in a successional series; it is self-perpetuating and in equilibrium with the physical
habitat. Both these definitions seem very similar and the differences do
not appear to be completely incompatible. However, they reflect the out-
look on climax vegetation of two schools of thought which have developed
since the initial usage of the word in an ecological context. The two theories
concerned are referred to as the monoclimax and the polyclimax theories
which respectively relate the control of climax vegetation to one factor,
that of climate, or to a number of factors ranging from climate to less
obvious edaphic or even biotic (i.e. animal) factors.

The existence of stable communities largely controlled by edaphic
factors are obviously not denied by the protagonists of the monoclimax
theory but they are regarded rather as 'exceptions' and categorized by the
introduction of special terms—'subclimax' and 'post-climax' for example
(see below). [Whitaker (1953) lists and describes in detail the numerous
terms used in relating edaphically controlled climax vegetation to the ultimate climatic climax.] It is claimed that over a long period of time these
apparent exceptions will be completely modified by the action of the
climate into true climax vegetation.

The polyclimax theory is not hampered by this necessity of introducing

a large number of terms to describe apparent exceptions to the theory—
an edaphic geological or physiographic climax is accepted for what it is.

The monoclimax theory was developed largely by Clements (1916) and
some of the terms relating to climax vegetation clearly show the inherent
difficulty of defining a time scale over which stability can be assessed.

Subclimax

This is the penultimate stage of a succession which persists for a long
time but is eventually replaced by the true climax vegetation. For example,
Gleason (1922) and Cooper (1926) give an account of the relic coniferous
forest uncovered in the city of Minneapolis and which was shown to be
immediately post-glacial. The present climax is deciduous forest which
presumably has very slowly replaced the coniferous 'subclimax'.

Disclimax

This is the vegetation replacing or modifying the true climax after a
disturbance of the environment. The majority of 'natural' grasslands in the
British Isles below 500 m are maintained by the grazing of cattle and
sheep, and if this factor was removed it is believed that a deciduous forest
climax would result. Similarly the introduction of prickly pear cactus in
Australia has formed a disclimax over wide areas.

Post-climax and Preclimax

Climatic areas, latitudinally zoned from the equator to the arctic, are
not discontinuous. There is a gradual change from one zone to the next
reflected in equally gradual vegetational change. Any slight fluctuation of
climate will modify the vegetation and such vegetational differences
reflecting cooler and/or moister conditions than the average (usually topographically induced) are termed post-climax, drier and/or hotter, pre-
climax.

The monoclimax theory and the excessive use of this type of terminology has been severely criticized by many ecologists since rather than
clarifying the position the dozens of special terms complicate the issue out
of all proportion. Conversely an explanation by polyclimax theory is
readily available.

The real difference between these two schools of thought lies not in the
possible existence of edaphically-stable communities but in the time factor
involved in any consideration of relative stability of vegetation. Selleck
(1960) expresses this very clearly: 'The rift (between the two schools of
thought) occurs either in the assumption that, given sufficient time,
climate is the over-all controlling factor of vegetation, or in the length of
time considered adequate for stabilization to occur.' In other words given
a sufficiently long period of time edaphic factors would be reduced to a

common level by the action of climate and the vegetation as a whole would develop uniformity accordingly.

It is impossible to define stability without reference to a time scale and since these are usually related to the life span of man himself, vegetational changes which occur over a 'geological time scale' are not readily appreciated. If the expectation of life was one minute then a population of groundsel (*Senecio vulgaris*) would be a climax vegetation. Conversely if our life expectation was measured in thousands of years the coniferous forest subclimax of North America would be merely a successional stage which precedes the establishment of deciduous forest. Such considerations are by no means as absurd as would be imagined. There is a continuous fluctuation of climate over large areas—the Pleistocene ice-age is an outstanding example—but other less marked changes have occurred in more recent times. Beschel (1961) gives data which shows that remarkable glacier fluctuations have occurred over the last 2000 years (Fig. 3.12). The evidence is taken from a variety of sources, geomorphology, historical records and dendrochronology, and shows a series of fluctuations consistent in areas as far apart as Greenland and Africa. The control of glacial activity is dependent on the over-all climate of the area and it is clear that there is a continuous variation of climate which can be very marked over a long period of time. Such climatic fluctuations will produce a successional response in apparently stable 'climax' vegetation. Cowles (1901) was well aware of this problem: 'The condition of equilibrium is never reached, and when we say that there is an approach to the mesophytic forest, we speak only roughly and approximately. As a matter of fact we have a variable approaching a variable rather than a constant.'

Whitaker (1953) attempted to circumvent the problem of time scale by proposing an approach based on the 'prevailing climax' which was related to the maximum area covered. As Selleck (1960) points out this can be criticized since a minor constituent of a forest understory for example could become the dominant species of the next generation.

One marked aspect of the theoretical climax relative to the stages of succession leading to it is the rate of change. The initial stages of succession are usually relatively rapid, the rate of change gradually decreasing until vegetational change is at a minimum and if we assume climatic and environmental stability all directional change ceases in the 'climax'. Again Cooper (1926), one of the early protagonists in this field, expressed this aspect very well: 'The climax period comes into being insensibly. Much,

---

Fig. 3.12. (*opposite*) Glacier advance and retreat in metres as summarized from lichenometrical, geomorphological, historical and dendrochronological evidence. (From Beschel 1961; courtesy of Univ. Toronto Press.)

Glacier advance ↑ and retreat ↓ in hypothermal time, from lichenometrical, geomorphological, historical. and dendrochronological evidence.

(1) Glacier Stanley occidental, Ruwenzori, Central Africa

(2) Ghiacciaio di Gran Neiron, Gran Paradiso, Italy

(3) Grünauferner, Stubai Alps, Tyrol

West Greenland (Locations of the glaciers are indicated on Figure 1.)

(4) Tasiussaq A

(5) Tasiussaq B

(6) Ikatussaq A

(7) Ikatussaq C

(8) Tunugdliarfik A

(9) Tunugdliarfik B

(10) Tunugdliarfik C

(11) Kugssuaq A

or even all, of a unit succession may be characterized by a gradual diminution of the rate of change. The climax itself being merely a continuation of this process, it is not possible to mark it off absolutely from the period of more active succession.' In general terms the rate of vegetational change would seem to offer a possible means by which climax vegetation could be delimited from the terminal stages of succession, but again the definition of the *relative* rates of change in the two appears to be insurmountable.

We are forced to conclude that climax vegetation is an abstract concept which is never reached in reality due to the continuous fluctuations of the climate (*directional* over periods of hundreds of years). Clearly the climate of a region has an over-all control of the vegetation. The regional development of tundra, coniferous forest, deciduous forest, etc., in a progression from the arctic towards the equator reflects this over-all control. However within each of these zones there exists a great variety of detailed modification, imposed by other factors such as geology, physiography and human activity. Such vegetation is equally stable over long periods of time and is self-perpetuating and on this basis the polyclimax theory is more sufficient than the monoclimax theory which represents an extreme abstract ideal. That the problem is relative, is inescapable and it is only possible to refer to vegetation as in an active process of successional development as compared to a state of extremely slow change, barely distinguishable from a self-perpetuating stable 'ideal'.

It is perhaps significant that the theory of climatic climax has been built up on the study of temperate vegetation and the evidence presented by Richards (1952) clearly indicates the widespread occurrence of stable vegetation types in the tropics determined by soil conditions. Thus Richards states, 'The primary forest types of Moraballi Creek, for instance, are all found under the same climate and there is no reason to consider any one of the five as less stable than the others. No process of development and no change of soil conditions is known or can be imagined which would, for example, convert the Wallaba forest on bleached sand into mixed forest on red loam or vice versa. The Mora forest, liable to flooding, shows no tendency to develop into mixed forest and the Mora soil could not conceivably develop into a mixed forest soil. It seems that all the Moraballi Creek primary types must be regarded as of equivalent status, as the Polyclimax theory demands. Each depends on a different, but equally permanent, combination of soil and topography developing under the same set of climatic conditions.' (See also Davis and Richards 1933, 1934.)

It is conceivable that climatic variation over long time scales is absent in the equatorial region, or is minimal relative to the marked glacier activity in the arctic and resultant plant successions outlined previously. No evidence is available which would demonstrate previous vegetational types in areas of present day Primary Rain forests and it is possible that the 'abstract' climax may in fact exist in such areas where there has been a minimum of climatic disturbance. On the other hand the fluctuation of climate has only recently been studied in detail and a decision as to the likelihood of major equatorial variations will have to wait further developments in meteorology. It is well known however that raised beaches occur around several of the large African lakes reflecting (presumably) different climatic conditions in the past.

One further aspect of the climax concept with profound implication in the delimitation of 'community units' is of importance: Clements regarded an association as a complex organism [or quasi-organism, Tansley (1920)] which arose through a variety of causes and subsequently grew, matured and died. In actively changing vegetation the association which replaces a given seral stage is different whereas in climax vegetation the 'death' of an association results in the establishment of a subsequent identical association. Implicit in this concept of the association as an organic entity or 'living organism' is the concept that climax associations can be readily recognized, *defined* and *classified*. The other extreme is presented by Gleason (1917, 1926) who proposed the individualistic concept of the plant association, which basically relates the continuously variable environment to an equally continuously variable vegetation. Thus no two associations are identical and implicit here is the fact that accordingly they cannot be classified and often cannot be clearly defined. Embodied in these two concepts are the two extremes of approach to community classification in use today. This will be considered in some detail at a later stage (see Chapter 9). In addition to this aspect of the climax association as a distinct and a recognizable entity Greig-Smith (1952*b*) points out: 'This (the complex organism of Clements) implies direct interaction between individuals both of the same and of different species.' Interaction between species or individuals of the same species can be conveniently measured by means of a χ^2-test or by an analysis of the 'pattern' of species distribution respectively (see Chapters 2 and 6). Greig-Smith investigated several sites in secondary forest ranging from areas recently disturbed (Site D) to areas disturbed many years before the investigation and only imperceptibly different from (undisturbed) primary forest (Site B). In addition an area of primary forest was sampled for comparison (Site E). The association between species is given in Tables 3.1, 3.2 and 3.3 as the numbers of observed and expected joint occurrences of pairs of species. Clearly the maximum amount of association is present in Site D (most recently disturbed) and least evidence of association occurs in Site E (primary forest). This is clear evidence that the maximum interaction between species inherent in the 'complex organistic' concept of climax vegetation is not valid at least in tropical rain forest. Greig-Smith

Table 3.1 Expected and observed numbers of joint occurrences of pairs of species in plots at site D (secondary forest). (From Greig-Smith 1952*b*; courtesy of *J. Ecol.*)

	Pisonia cuspidata 73	*Rudgea freemani* 72	*Amaioua corymbosa* 58	*Miconia prasina* 58	*Lacistema aggregatum* 54	*Alibertia acuminata* 42	*Protium guianense* 42	*Coccoloba latifolia* 35
Rudgea freemani 72	52·56 / 53							
Amaioua corymbosa 58	42·34 / 38*	41·76 / 44						
Miconia prasina 58	42·34 / 34***	41·76 / 39	33·64 / 24***					
Lacistema aggregatum 54	39·42 / 48	38·88 / 43	31·32 / 37*	31·32 / 26*				
Alibertia acuminata 42	30·66 / 26*	30·24 / 33	24·36 / 29	24·36 / 19*	22·68 / 29*			
Protium guianense 42	30·66 / 32	30·24 / 34	24·36 / 26	24·36 / 22	22·68 / 30**	17·64 / 17		
Coccoloba latifolia 35	25·55 / 27	25·20 / 23	20·30 / 25*	20·30 / 13*	18·90 / 21	14·70 / 11	14·70 / 13	
Vismia guianensis 23	16·79 / 13*	16·56 / 16	13·34 / 15	13·34 / 10	12·42 / 10	9·66 / 12	9·66 / 8	8·05 / 7
Bactris sp. 20	14·60 / 17	14·40 / 17	11·60 / 11	11·60 / 11	10·80 / 12	8·40 / 7	8·40 / 12	7·00 / 7
Isertia parviflora 19	13·87 / 14	13·68 / 13	11·02 / 10	11·02 / 11	10·26 / 6*	7·98 / 6	7·98 / 8	6·65 / 4
Clusia minor 14	– / –	– / –	8·12 / 9	8·12 / 8	7·56 / 7	5·88 / 5	5·88 / 5	
Cassia fruticosa 13	– / –	– / –	7·54 / 8	7·54 / 5	7·02 / 11*	5·46 / 5	5·46 / 7	
Casearia spinescens 13	– / –	– / –	7·54 / 8	7·54 / 6	7·02 / 10	5·46 / 8	5·46 / 7	

The number against each species is the total number of plots in which that species occurred. The numbers against each pair of species are the expected (above) and observed (below) numbers of joint occurrences.

$* P = 0.01{-}0.05.$ $** P = 0.001{-}0.01.$ $*** P = {<}0.001.$

thus concludes that the individualistic concept of Gleason is a more satisfactory alternative in part at least. The evidence relating the pattern of distribution of the individuals of a species in an area to the 'stability' of the vegetation, suggests equally a minimum of interaction in stable vegetation. Greig-Smith (1961) examined the different pattern scales in *Ammophila arenaria*, sampling young developing dunes, stable dunes and dune 'slacks'. His data (Fig. 3.13) show almost complete absence of pattern in the early formation of dunes (NAD 12), a maximum of pattern in the stabilizing dunes (NAD 7, 8 and 9) and only slight indications of pattern in the most stable sites examined (NAD 2, 3 and 4). The amplitude of the peaks is proportional to the intensity of the patterns. In this example the pattern varies from one consisting of patches of slightly higher density which alternate with similar patches of lower density, to sites where the patches are more aptly described as 'clumps'. Again the evidence suggests

Table 3.2   Expected and observed numbers of joint occurrences of pairs of species in plots at site B (secondary forest). (From Greig-Smith 1952b; courtesy of *J. Ecol.*)

	Cordia curassavica 60	*Rudgea freemani* 50	*Casearia spinescens* 45	*Bactris sp.* (stems) 35	*Fagara martinicensis* 33	*Bactris sp.* (seedlings) 32
Rudgea freemani 50	30·00 / 34					
Casearia spinescens 45	27·00 / 31	22·50 / 30**				
Bactris sp. (stems) 35	21·00 / 21	17·50 / 18	15·75 / 11*			
Fagara martinicensis 33	19·80 / 21	16·50 / 15	14·85 / 18	11·55 / 10		
Bactris sp. (seedlings) 32	19·20 / 17	16·00 / 14	14·40 / 16	11·20 / 12	10·56 / 10	
Acnistus arborescens 31	18·60 / 21	15·50 / 15	13·95 / 14	10·85 / 13	10·23 / 14	9·92 / 10
Casearia decandra 14	18·40 / 8	7·00 / 9	6·30 / 13***			
Casearia guianensis 13	7·80 / 10	6·50 / 9	5·85 / 10**			

The number against each species is the total number of plots in which that species occurred. The numbers against each pair of species are the expected (above) and observed (below) numbers of joint occurrences.   * $P = 0\cdot01\text{--}0\cdot05$.   ** $P = 0\cdot001\text{--}0\cdot01$.   *** $P = <0\cdot001$.

random distribution in the final stable vegetation, that is to say no inter-action between individuals of a species.

A similar set of data is given by Kershaw (1963) collected from associations of *Calamagrostis neglecta* in Central Iceland. It behaves here as a primary coloniser of open water and remains at a reduced level of performance and density while the silting up of the area continues. The recent volcanic nature of the rock in the area results in the annual deposition of vast quantities of cindery silt on the valley floors and by an inspection of the depth of water table below a stand of *Calamagrostis* it is possible to arrange a relative time scale for each stand. The deeper the water table and the more the silting, the longer the *Calamagrostis* has maintained itself in that particular site. In detail such a method of dating would be open to severe criticism but it does give an indication of the relative age of stands. Samples were taken from numerous sites. The data (Fig. 3.14) show an identical trend. There is a minimal pattern intensity in the early stages of *Calamagrostis* establishment (Fig. 3.14a) followed by an increased intensity until silt accumulation lowers the water table to about 25 cm (Fig. 3.14b and c). At about 50 cm the pattern of *Calamagrostis* though still clear is very much reduced in intensity. This trend is continued (Fig. 3.14d–g) in the series until the maximum depth of water table

Table 3.3 Expected and observed numbers of joint occurrences of pairs of species in plots at site E (primary forest). (From Greig-Smith 1952b; courtesy of J. Ecol.)

	Myrcia granulata 59	Rudgea freemani 45	Hymenaea courbaril 37	Rinorea lindeniana 33	Eugenia perplexans 26
Rudgea freemani 45	26·55 / 30				
Hymenaea courbaril 37	21·83 / 25	16·65 / 17			
Rinorea lindeniana 33	19·47 / 17	14·85 / 20*	12·21 / 7*		
Eugenia perplexans 26	15·34 / 17	11·70 / 15	9·62 / 12	8·58 / 10	
Peltogyne porphyrocardia 24	14·16 / 16	10·80 / 8	8·88 / 6	7·92 / 5	6·24 / 7
Swartzia simplex 20	11·80 / 11	9·00 / 8	7·40 / 10	6·60 / 3	5·20 / 5
Calptranthes fasciculata 16	9·44 / 8	7·20 / 8	5·92 / 3	5·28 / 7	
Clathrotropis brachypetala 16	9·44 / 8	7·20 / 6	5·92 / 4	5·28 / 3	
Brownea latifolia 16	9·44 / 7	7·20 / 6	5·92 / 6	5·28 / 5	
Eschweilera subglandulosa 15	8·85 / 9	6·75 / 9	5·55 / 7		
Bactris sp. 14	8·26 / 10	6·30 / 7	5·18 / 7		

The number against each species is the total number of plots in which that species occurred. The numbers against each pair of species are the expected (above) and observed (below) numbers of joint occurrences.

$* P = 0·01–0·05.$

underlying *Calamagrostis* is reached at about $1\frac{1}{2}$ m in the 'desert areas', after which most vegetation disappears altogether. A further collection of data was made in the area in a topographically uniform *Rhacomitrium* heath situated at about 790 m on the flat summit of a mountain. From the remoteness and past history of the region it would appear that the area had been completely undisturbed and from the evidence outlined above little pattern should have been present, apart from that produced by the morphology of the species. Again the analysis of the data (Fig. 3.15) is consistent with the concept of a minimum of pattern (interaction between individuals) in stable vegetation. The pattern detected is due to the morphology of the plant and is equivalent to the dimensions of the area covered by the rhizome system of an individual plant.

Brereton (1971) examining scales of pattern and their intensity (measured as peak height) showed for two salt marsh species *Puccinellia* and *Salicornia* a marked change in both pattern scale as well as pattern intensity. The sites sampled ranged from high water mark spring tides

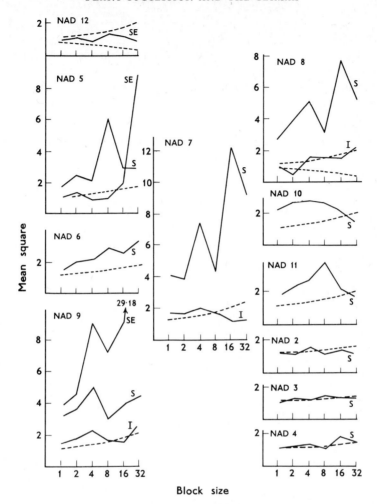

Fig. 3.13. The changing intensities of pattern of *Ammophila arenaria* in relation to dune age and stability. The dotted lines represent upper and lower 95 per cent limits for random distribution. (From Greig-Smith 1961; courtesy of *J. Ecol.*)

(stand 1) successively down through the marsh, to the lower limit of *Salicornia* at approximately the level of high water at ordinary neap tides (stand 6). The results (Fig. 3.16) are extremely clear cut and call for little comment.

The evidence is consistent; it is drawn from a variety of habitats ranging from sub-arctic moss-heath to tropical forest and indicates that the original thesis of Greig-Smith's is correct.

Conclusions

The controversy centred on the climax concept has been continuous for almost 50 years, the literature is bulky and has been condensed here to a considerable extent. It may help to clarify the position by summarizing the more important aspects:

1. Succession is defined as a directional vegetational change induced by an environmental change or by intrinsic properties of the plants.

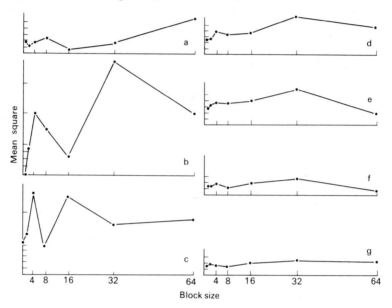

Fig. 3.14. The changing intensities of pattern of *Calamagrostis neglecta* in relation to age and stability. (From Kershaw 1963; courtesy of *Ecology*.)

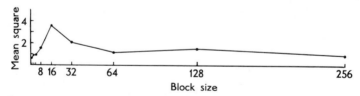

Fig. 3.15. General absence of pattern on *Carex bigelowii* growing in stable *Rhacomitrium* heath other than small scale morphological pattern (see text). (From Kershaw 1963; courtesy of *Ecology*.)

2. The initial rate of vegetational change is high and subsequently falls to a low level after which further development is governed by very slow changes of climate or physiography. This relatively stable state is referred to as the climax.

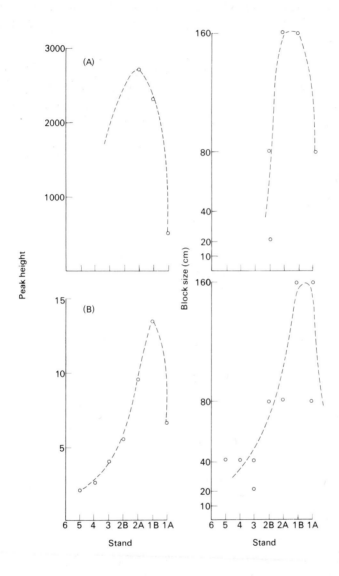

Fig. 3.16. The change in intensity and pattern scale in a salt marsh succession, from high water (1A) to low water mark (6). (A) *Puccinellia*. (B) *Salicornia*. (From Brereton 1971; courtesy of *J. Ecol.*)

3. The original 'climatic climax' is an abstract concept since climate is *not* stable and furthermore other environmental factors influence the final vegetation of an area in addition to climate. Thus the polyclimax theory is a more realistic concept.

4. Climax vegetation is defined approximately as a self-perpetuating association of plants having a maximum of stability; any change occurring is very slow, and only appreciable over extensive time scales (relative to the time scale of establishment of the climax through succession).

5. Evidence shows stable vegetation to have a minimum of association between species and individuals (pattern). Thus Gleason's individualistic concept of the climax is a more satisfactory hypothesis than the 'complex organism' as envisaged by Clements.

4 Cyclic Vegetational Change

IT is commonplace to regard vegetation which is stable and in equilibrium with its environment as a static entity, which having gone through various stages of succession has finally reached an ultimate level of stability and ceases to change any further. This superficial impression is misleading since in addition to the external environmental factors which regulate the over-all composition of the vegetation, there exists a series of intrinsic factors which continue to induce vegetational change. This intrinsic vegetational change is not manifested as a continuation of succession but as a cycle of events at any particular point in a community. These events are replicated many times over the whole of the community, and are apparent as a series of phases present at a number of points at any instant in time. The over-all summation of these phases constitutes the composition of the community. Thus while plant succession is a *directional change*, cyclic changes represent *fluctuations about a mean value*.

HUMMOCK AND HOLLOW CYCLE

Watt (1947*a*) outlined a number of examples of cyclic fluctuations containing several marked phases which are now regarded as classic examples of cyclic change. The regeneration complex which occurs on actively growing raised bog (the 'hummock and hollow cycle' as it is often called) is an outstanding example of a series of dynamic phases which are related to each other in time. The phases form a mosaic of patches, each at a different stage from its neighbours. Oswald (1923) and Godwin and Conway (1939) have shown the relation of the phases to each other by examination of the stratigraphy of the underlying peat. The initial phase is represented by *Sphagnum cuspidatum* which invades small pools of water. This species is invaded by *S. pulchrum* which in turn is replaced by *S. papillosum*. The latter species build up low hummocks which are eventually colonized by *Calluna vulgaris*, *Erica tetralix*, *Eriophorum vaginatum* and *Trichophorum caespitosum*. At a later stage the lichen species *Cladonia arbuscula* (*C. sylvatica*) enters the sequence, each sequence being abundantly preserved as stratified deposits of peat. At this stage the sequence is identical to a 'micro-succession' and each of the subsequent stages is dependent on the reactions on the environment by its predecessors. However, as the *Calluna* ages and dies the hummock is eroded and a pool of open water re-established largely by the activity of the adjacent phases building up hummocks. Thus, under these conditions, an area of raised bog is a mosaic of patches each of which represents a phase of a cycle; each patch will progress eventually and repeatedly through the same cycle of events and although a given (small) area will change over a period of time the whole community remains essentially the same.

A similar example is drawn by Watt from grassland, where the vegetational variation is related to microtopography. Four phases are present, descriptively termed 'hollow', 'building', 'mature' and 'degenerate' (Watt 1940). The hollows are invaded by seedlings of *Festuca ovina* around which accumulates a variety of wind-borne particles in addition to humus remains of the parent plant. Over a period of time a small hummock is formed, representing the 'building' phase (Fig. 4.1). In the mature phase

Fig. 4.1. Diagrammatic representation of the *Festuca ovina* erosion cycle indicating 'fossil' shoot bases in the soil. (From Watt 1947; courtesy of *J. Ecol.*)

the hummocks reach a height of about 4 cm and are colonized by several lichen species notably *Cladonia alcicornis* and *C. rangiformis*. The lichen species slowly assume a more dominant role in the early stages of the 'degenerate' phase and are gradually replaced by crustaceous lichen species. Erosion of the hummock follows completing the full cycle of events back to the 'hollow' phase. Again the over-all structure and composition of the community remains unchanged since each phase is represented by a mosaic of cyclic units related spatially to each other as well as representing a time sequence.

The occurrence of cyclical phases is not restricted to perennial herbaceous species and Watt (1947) also describes a cyclical development which occurs over a long period in beech woodland on chalk plateau. Following the death of a full grown tree the gap that is left is initially devoid of vegetation and represents the first phase of the cycle. This is invaded initially by *Oxalis* and subsequently by *Rubus*, representing the next two phases of the cycle. Concurrently beech and ash seedlings become established forming a 'reproduction circle' (Watt 1925) with the ash in the centre surrounded by a circle of young beech originating from seed shed by the surrounding mature beech. Eventually the ash is over-topped by the beech saplings and the gap finally closed. The cycle is finally completed by the ultimate degeneration and death of the beech colonizers initiating the same cycle again. The time scale in this example is very long but the nature of the process is identical.

In general the sequence of events is similar in all the examples given by Watt (1947a) and these he descriptively terms pioneer, building, mature and degenerate. The productivity of each phase appears to follow a generalized curve of increasing productivity in the early phases of a cycle, leading to an optimum in the mature phase, and followed by reduced productivity in the degenerate phase (Fig. 4.2). Billings and Mooney (1959) give an account of an extremely interesting cycle in tundra vegetation at 3600 m in the Medicine Bow Mountains of southern Wyoming. The area as a whole is subjected to intense soil frost action and the valley bottom is covered with actively forming frost hummocks of peat and stone polygons. The occurrence of frost hummocks and polygons is extremely characteristic of arctic and sub-arctic zones, where intense frost produces a number of marked topographical features. Billings and Mooney relate the formation of polygons in this area to the degradation of the frost hummocks of peat. When a hummock rises to a certain level so that it protrudes above the general level of the snow in winter, the species on the crest of the hummock are killed and eroded away by the repeated action of wind and wind-borne ice/snow particles. With continued exposure the hummock itself is eroded exposing rocks and silt which are pushed up concurrently from below by further frost action. Eventually the hummock is completely eroded to its base and this stage appears as a stone polygon surrounded with a 'rim' of peat.

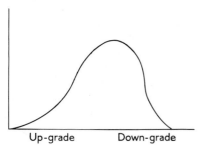

Fig. 4.2. Generalized productivity curve. (Redrawn from Watt 1947a.)

Up-grade   Down-grade

The water table in the whole area fluctuates from year to year; the hollows usually are fairly wet and often whole areas of polygons are completely submerged. During these submerged periods the hummock formation appears to be initiated by *Carex aquatilis* which invades the rocky intervals of the polygons. The general sequence of events is shown in Fig. 4.3 below. Following the establishment of *C. aquatilis, Sedum*

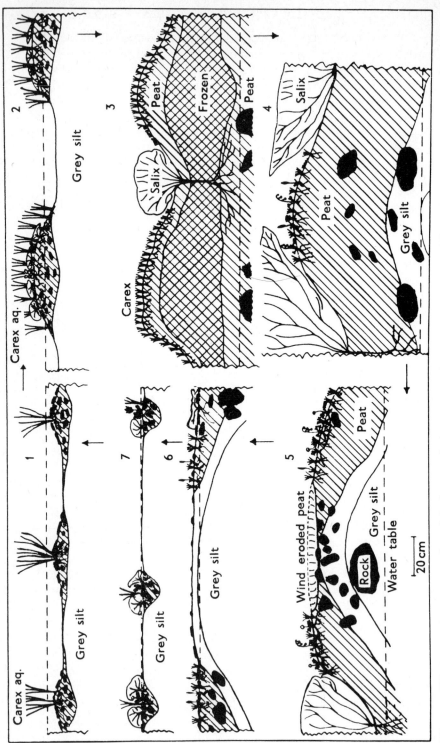

Fig. 4.3: Diagrammatic representation of the hummock, frost scar and polygon cycle. (From Billings and Mark 1961; courtesy of *Ecology*.)

rhodanthum and several bryophyte species become established. Eventually these species are replaced by *Geum turbinatum* and *Polygonum viviparum* which form a closed turf with other additional species on the mature hummocks. The erosion commences at this stage of the cycle and subsequent drowning of polygons initiates the next cycle. Thus there is a continuous cycle of change, hummocks alternating with the centre of polygons and the margin of the polygons alternating with subsequent hummocks. The general phasic development of this cycle is strikingly similar to the hummock and hollow cycles discussed by Watt (1947a). The mechanism is partly different and the species involved obviously different, but the same general principle is common to all three cases. The pioneer phase in this instance is dominated by *Carex aquatilis*, the building phase by *Carex* and *Sedum* and the mature phase by *Geum* and *Polygonum*.

MARGINAL EFFECTS

Cyclic phases as described above are partly dependent on intrinsic properties of the plants themselves and partly dependent on intrinsic properties of the community both of which are modified and controlled to a greater or lesser extent by the environment. Thus the ability of *Sphagnum papillosum* to form hummocks is an intrinsic property of that species, the availability of the other species involved in the cycle is an intrinsic property of the community, and the final partial erosion of the hummock is a result of environmental control. The downgrade phases of the cycle are terminated by erosion of the hummock but it is interesting to speculate as to the actual initiation of the downgrade phase. A possible explanation lies in the relationship between the general vigour of a perennial plant and its age. This important concept was originally outlined by Watt (1947b) in relation to the vigour of bracken invading grassland where the marginal zone was considerably taller and apparently more vigorous than in the 'hinterland'. The difference in height of the fronds in the margin and hinterland is considerable (Fig. 4.4). The immediate possibility which would account for this 'marginal effect' would be the presence of some factor necessary for the full development of the fronds which is markedly reduced in quantity by the advancing front and is in short supply in the hinterland. However, analysis of the soil (Table 4.1) shows no consistent trend to support such a hypothesis and furthermore the addition of different combinations of fertilizers containing nitrogen, potash and phosphorus to sample plots in the hinterland had no significant effect.

The significant feature is the progressive increase of rhizome age back from the margin to the hinterland where the oldest part of the rhizome decays and lateral shoots become separated as individuals. Thus the margin has a large number of young even-aged rhizomes orientated parallel to

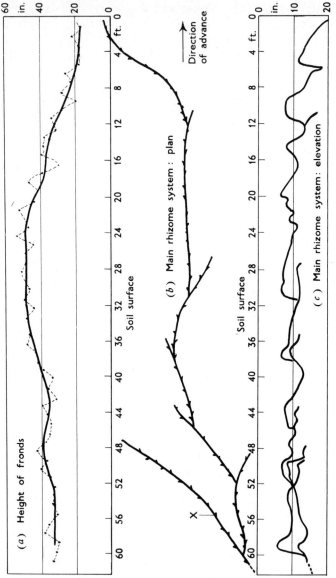

Fig. 4.4. (*a*) the average height of fronds across the advancing margin of bracken; (*b*) plan and (*c*) elevation of main rhizome. (From Watt 1940; courtesy of *New Phytol.*)

each other. These younger parts of the rhizome produce larger and more vigorous shoots resulting in the marked marginal effect. Conversely in the hinterland the rhizomes are initially much older and the shoots smaller. As the main rhizome decays lateral shoots of different ages are separated as individuals each of which will subsequently follow the same progressive change of vigour as each individual ages. Thus the hinterland consists of

Table 4.1—Soil data from surface layers (0–7·6 cm) in different zones of bracken invading grassland. (From Watt 1940; courtesy of *New Phytol.*)

	In grassland	In zone of maxi-mum height	In dense uniform bracken	In hinterland
pH	4·2	4·0	3·8	4·0
Carbon per cent	1·785	1·425	1·740	1·610
Carbon/Nitrogen	19·40	19·79	19·12	19·63
Exchangeable Calcium in marginal effect	1·44	1·28	1·28	1·36
Total exchangeable bases	2·40	2·00	1·61	1·61

very unevenly aged groups of individuals at all stages of the cycle forming a mosaic of phases. In addition to the height of frond being at its maximum in the mature phase the length of petiole above ground follows the same trend and the length of petiole below ground level follows an inverse trend (Fig. 4.5). There is a marked relationship between the age of the

Fig. 4.5. The positions and dimensions of bracken fronds showing the sequence of phases from pioneer to degenerate. The dots indicate the depth of origin, petiole length and height of fronds. (From Watt 1947*b*; courtesy of *New Phytol.*)

rhizome and its morphology, and also between the vigour or performance of the aerial shoots from young and older parts of the same rhizome. This feature may explain the onset of the downgrade stage of the cyclic phases outlined above. As the principal species of the mature phase age, their performance falls markedly and the degenerate phase commences.

The general occurrence of a phasic development of perennial plants is indicated from the numerous reports of marginal effects in the literature. Kershaw (1962*a*) describes a very marked marginal effect in *Eriophorum angustifolium* invading a *Salix herbacea/Polygonum viviparum* community in central Iceland. The density and performance (expressed as leaf length, leaf width and length of leaf tip) both show a similar relationship to the advancing front in *Pteridium* described by Watt (Figs. 4.6, 4.7 and 4.8).

Fig. 4.6. The variation of performance of *Eriophorum* (leaf length) along a transect running through the circular advancing margin. (From Kershaw 1962; courtesy of *J. Ecol.*)

Fig. 4.7. The variation of performance of *Eriophorum* (leaf width) along a transect running through the circular advancing margin. (From Kershaw 1962; courtesy of *J. Ecol.*)

Similarly Ovington (1953), Caldwell (1957) and Penzes (1958, 1960) have described numerous examples of advancing fronts of even-aged rhizomes at an optimum level of performance relative to the hinterland with its very much lower level. Watt (1955) shows that *Calluna* follows the same phase development and furthermore where *Pteridium* is present there is an inverse relationship between the two species. Where *Calluna* is at its building/mature phase, competing bracken is at its minimum development. Similarly where *Calluna* is in its degenerate phase, bracken is at the

Fig. 4.8. The variation of performance of *Eriophorum* (tip length) along a transect running through the circular advancing margin. The length of tip is inversely proportional to performance. (From Kershaw 1962; courtesy of *J. Ecol.*)

building or mature phase. Thus concurrently the competitive ability changes with the phasic series pioneer, building, mature and degenerate. The phasic development of *Pteridium* and *Calluna* have been confirmed by Anderson (1961a, 1961b) and Nicholson and Robertson (1958).

INTERACTION OF AGE AND PERFORMANCE

Confirmation of the relationship between age of an individual and its competitive ability is given by Kershaw (1962). The effect of the advancing front of *Eriophorum* (see p. 72) was related to the density of *Salix herbacea*, the density of *Salix* being inversely related to the density and performance of *Eriophorum* (Figs. 4.9 and 4.10). In effect where the density of *Salix* was at a minimum *Eriophorum* had a maximum level of performance and abundance, indicating the greater competitive ability of

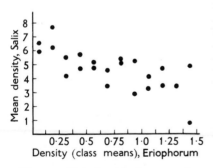

Fig. 4.9. Relationship between the densities of *Eriophorum* and *Salix herbacea* (see text). (From Kershaw 1962; courtesy of *J. Ecol.*)

Fig. 4.10. Relationship between the density of *Salix herbacea* and performance (leaf length) of *Eriophorum* (see text). (From Kershaw 1962; courtesy of *J. Ecol.*)

Eriophorum in its building and mature phases. Thus in general terms there exists a direct relationship between age, performance and competitive ability, potentially, for most if not all perennial plants.

The quantitative assessment of the age/performance relation presents some difficulties since it is often impossible to determine the age of most herbaceous perennials. Data for two species only are available (Kershaw 1960b, 1962c), for *Alchemilla alpina* and *Carex bigelowii* the assessment of age being based on ring counts and the rhizome morphology respectively. The mean leaf diameter and leaf number per individual of *Alchemilla* were used as a measure of performance, each individual being aged by a count of growth rings. A considerable variation of performance is present (Fig. 4.11) between individuals of the same age reflecting local environ-

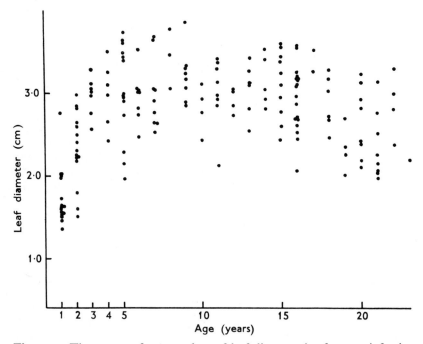

Fig. 4.11. The scatter of mean values of leaf diameter (performance) for individual plants of *Alchemilla alpina* of all ages. (From Kershaw 1960; courtesy of *J. Ecol.*)

mental effects, but the total data shows a clear rise and fall in the relationship between performance and age. A regression of the form $Y = a + bx + cx^2$ is highly significant and indicates the mature phase to fall within the age groups 9–15 years (Fig. 4.12). The relationship between leaf number and age is similar but with a larger scatter of values (Fig. 4.13). The data

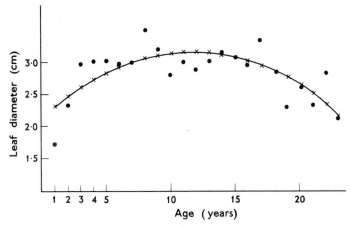

Fig. 4.12. Fitted curve of the type $Y = a + bx + cx^2$ for the total leaf diameter/age data of *Alchemilla*. (From Kershaw 1960; courtesy of *J. Ecol.*)

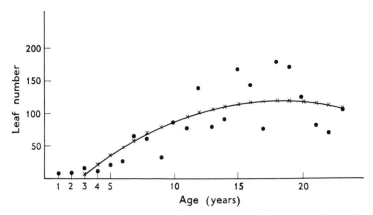

Fig. 4.13. Fitted curve of the type $Y = a + bx + cx^2$ for the total number of leaves/age data of *Alchemilla*. (From Kershaw 1960; courtesy of *J. Ecol.*)

clearly confirm the relationship initially proposed by Watt to describe the phasic development of *Pteridium*. The data for *Carex bigelowii* is similar (Fig. 4.14). Leaf length and leaf width are used as a measure of the performance and the age of a rhizome system determined by careful morphological examination. At the end of a season's growth, the aerial parts die, the basal leaf sheaths remain as a humified 'knot' on the rhizome and it is relatively straightforward to determine the age of a rhizome system with a fair degree of accuracy.

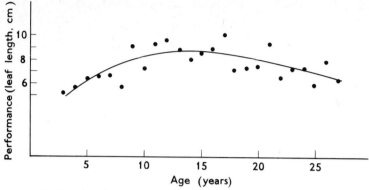

Fig. 4.14. Performance/age relationship of *Carex bigelowii* as expressed by leaf length. A curve of the type $Y = a + bx + cx^2$ is fitted. (From Kershaw 1962; courtesy of *J. Ecol.*)

It seems highly probable that similar growth curves will exist for many other perennial species. For this reason phasic development and relationships between other species are likely to be widespread and the lack of additional evidence reflects the difficulties of detecting such cyclic relationships. Phasic development involving several species in a definite sequence seems unlikely to be as widespread as a mosaic of density phases reflecting the age/performance interaction with the attendant variation of competitive ability and the potential effects on associated species. Thus a stable community will be in a constant state of phasic fluctuation, one species becoming locally more abundant as another species reaches its degenerate phase, the fluctuation in density being however slight and not visually distinct. Only by careful quantitative measurement of abundance and performance can phasic developments be detected in vegetation where there is a complete cover, and then only in those species which possess some morphological or anatomical feature which enables their age to be assessed. Where the vegetational cover is discontinuous, and the degenerate stage is eroded and followed by a 'bare' phase, cyclic relationships become visually more obvious but apparently rather uncommon. However, it is clear that this form of relationship is likely to be widespread and is the mechanism underlying the dynamic nature of vegetation. The initial demonstration of non-randomness in vegetation was somewhat surprising to the early workers in this field and they were at a loss to account for it apart from consideration of morphology and localized seed dispersal (see Chapter 8, p. 145). In the light of the evidence outlined above it is clear that a non-random distribution of individuals will always occur however uniform the environment, as long as several age groups are present.

Such patterns have been detected by Kershaw (1958, 1962) in *Agrostis tenuis, Calamagrostis neglecta* and *Carex bigelowii* where small clumps of tillers a few centimetres in diameter are formed from different rhizome systems (p. 164). A typical turf from upland rough grazing in Wales consisted of a mosaic of *Agrostis* tillers surrounded by *Festuca* and various bryophyte species, and in all cases the groups of *Agrostis* shoots arose from several different rhizome systems (Figs. 8.15 and 8.16). The mechanism of this rhizome orientation is somewhat obscure but if any one group of tillers is considered in isolation over a period of time it follows a typical phasic development. Initially a group of young tillers is established from a few pioneer rhizomes, representing the pioneer stage. The following year the clump produces several axillary shoots and is enlarged by the arrival of further pioneer rhizomes; the building phase. In subsequent years this process is continued through a mature phase until there is a preponderance of dead tiller bases in the clump and the degenerate phase follows with increasing numbers of dead tiller bases and weaker aerial shoots. Finally the whole phase is represented merely by humified remains of shoot bases which can be found underlying a small patch of *Festuca*. By careful examination the different phases can be picked out by the size and colour of the leaves of *Agrostis*. The early phases have bright green small leaves, the building phase bright green and larger leaves with little evidence of dead leaf bases. The mature phase though similar has a number of dead leaf bases present, a trend which becomes increasingly evident in the degenerate phase where there is a preponderance of dead leaves with a few small and often yellowish-green aerial leaves.

The underlying mechanism of these varying levels of performance of individual plants and the corresponding relationship of their competitive ability has been obscure. Recently however a fuller understanding of the competition for light by crop plants has been achieved (Donald 1961) and these results seem to be of direct significance in this present context. If water and nutrients are in adequate supply, then light will become the limiting factor controlling rate of growth and the production of dry matter. In any stratified vegetation there exists a very sharp decline in light intensity from the surface layers downwards and very often the lowest leaves will make a negative contribution to the net assimilation of the whole 'canopy' since the light intensity will be insufficient to keep them above compensation point. The respiration of these lower leaves in fact exceeds their photosynthesis. Expressing leaf area as a leaf area index (i.e. leaf area per unit of land area), it can be shown theoretically that if the relationship of photosynthesis to respiration is extrapolated for a continued increase of leaf area index, then a leaf area index will be reached at which the respiration of the whole canopy of foliage will equal its photosynthesis. Under natural conditions there is little evidence to suggest that this state is ever achieved in crop plants, but it seems more likely that perennial plants may reach this compensation point. From an ecological point of view it is this possibility which is of significance, since not only are the aerial parts of importance but the underground parts also. Thus a perennial plant with an extensive rhizome system has a large actively respiring volume with a relatively small (theoretically peripheral) zone of aerial shoots, and the compensation point is likely to be reached or exceeded in old individuals. *Carex bigelowii* is a case in point. A complete rhizome system 20 years of age will cover one square metre with a complex system of branches, each ultimate branch having a terminal aerial shoot and only one or two axillary shoots in the last two years' growth increment. In such a system it is very likely that the compensation point has been reached or exceeded and the plant is in its 'degenerate phase'. Conversely a young plant of 6–12 years of age has a much less extensive rhizome system and with a normally developed 'leaf canopy' is at its optimum level of growth.

Went (1957) has referred to the build-up of non-photosynthesizing tissues in old strawberry plants until the whole plant is barely at compensation point at a high light intensity and it seems probable that the ratio between the photosynthesis and respiration of a perennial plant largely controls its general level of performance and its ability to compete with its neighbours. This possible explanation is equally satisfactory when applied to the parts of a fragmented degenerate individual. In *Alchemilla*, *Agrostis*, *Pteridium* and *Carex* the older parts of the rhizome slowly decay and as the zone of decay reaches a lateral branch this becomes separated off as an individual. The oldest plants of low performance thus release 'propagules' with, initially at least, a low performance but with a morphologically optimal age. A propagule from a 25-year-old parent is likely to be 8–12 years of age. However the average performance of all individuals in this age group is optimal and accordingly it would appear that the level of performance of such a propagule immediately rises to its 'morphological age'. This seems quite logical when considered as a small rhizome system being severed from a very large system which was respiring actively. Immediately the photosynthesis/respiration ratio becomes unequal, the terminal leaves provide an excess of metabolites relative to the now small rhizome system, and active growth can take place at once.

In all the cycles described above, the species involved are in one sense, part of a truncated succession. There is no directional change, the cycle eventually returning to its original position. However, just as in plant succession where there is a marked reaction of the species on the environment, there is a similar reaction on the environment throughout the cycle of phases. Furthermore, because of the phasic sequence the level of reaction will also vary sequentially. Barclay-Estrup (1971) examined the microclimate and its variation in relation to the phasic development of *Calluna vulgaris* (Watt 1955), and showed the existence of considerable differences between the different phases. Soil surface temperatures during the summer months were much higher under the building phase than elsewhere and followed a decreasing sequence through the building phase and mature phase to the degenerate phase (Fig. 4.15). Conversely in the winter months the lowest temperatures recorded are under the pioneer and degenerate phases where the low biomass of the *Calluna* allows a greater loss of radiant energy from the soil. Similar considerations apply

Fig. 4.15. Surface maximum and minimum temperatures for the period January 1964–January 1966. (a) Mean maximum temperatures; (b) Mean minimum temperatures. —○—, Pioneer; —×—, building; – – ○ – –, mature; – – × – –, degenerate. (From Barclay–Estrup 1971; courtesy of *J. Ecol.*)

to soil temperature although the differences are not as clear cut. Interception of light is also related to the phasic sequence and during the building/mature phases, the light at the soil surface falls to 10-20 per cent of full illumination, as opposed to 60-80 per cent in the pioneer phase, and 40-60 per cent in the degenerate phase (Fig. 4.16). Rainfall interception follows the same sequence with maximum interception by the canopy of the mature and building phases, decreasing through the degenerate phase to the pioneer phase where cumulative precipitation is fairly close to full precipitation measured in the open (Fig. 4.17).

Fig. 4.16. Means of all ground-level photocell measurements. Each graph shows the light values along two lines crossing the quadrant in a south–north direction (○) and an east–west direction (×). (From Barclay-Estrup 1971; courtesy of *J. Ecol.*)

Clearly the microclimate at the soil surface varies markedly over a period of time in response to the phasic changes in the *Calluna*. In response to this mosaic of microclimate there is a development of corresponding

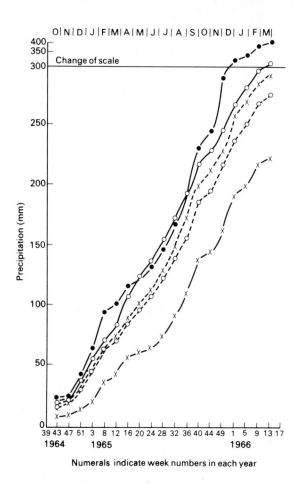

Fig. 4.17. Cumulative microprecipitation under *Calluna* of each phase. —○—, Pioneer; —×—, building; – – ○ – –, mature; – – × – –, degenerate; —●—, open. (From Barclay-Estrup 1971; courtesy of *J. Ecol.*)

mosaics of abundance levels in associated species (Fig. 4.18). Notably *Vaccinum vitis-idaea*, *V. myrtillus*, *Empetrum nigrum*, and the moss species *Polytrichum commune*, *Pleurozium schreberi* and *Hylocomium splendens* dominate the more open areas during the pioneer phase, disappearing during the building and mature phases, and reappearing again in the

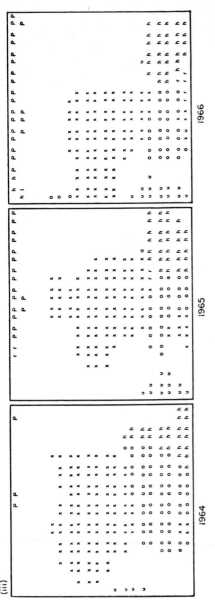

Fig. 4.18. Periodic maps of quadrat 8, containing *Calluna* in the degenerate phase. (i) *Calluna*: hatched areas, adjacent building plant; unshaded outlined areas, pioneer plant; outlined area (broken line), gap boundary. (ii) Other vascular plants. (iii) Bryophytes, C, *Calluna*; d, *Deschampsia flexuosa*; h, *Plagiothecium undulatum*; m, *Vaccinium vitis-idaea*; o, *Hylocomium splendens*; p, *Hypnum cupressiforme*; r, *Aulacomnium palustre*; u, *Pleurozium schreberi*; v, *Empetrum nigrum*; x, *Polytrichum commune*. (From Barclay – Estrup 1969; courtesy of *J. Ecol.*)

degenerate phase. These results amplify the previous discussion on cyclic phases as a causal mechanism of pattern in vegetation. Thus the phasic development of *Calluna* affects other associated species through its changing levels of competitive ability. This is evident as a direct interaction with the environment inducing a change in the microclimate. The resultant spatial patterns of the bryophyte species especially are controlled by this changing microclimate.

5 Correlation and the Causal Factors of Positive and Negative Association between Species

CORRELATION COEFFICIENTS

THE detection of association by means of a contingency table and calculation of χ^2, implies a proportion of quadrats which do not contain the species under consideration. An important aspect of community structure is the relationship between the *quantities* of species present, rather than the mere presence or absence of the species. Two species in fact may show a marked negative *correlation*, the amount of one species varying inversely with that of the other even in cases when all the quadrats of the sample contain at least one individual of each species, and hence cannot be analysed by the χ^2 method. Where the data are in the form of cover, frequency or density, a correlation coefficient is the simplest measure which detects any possible relationship between them. Similarly relationships between abundance of a species and measurements of the environment can be detected by correlation coefficients. A similar approach is to calculate the equation which best fits the data, or in other words to calculate the 'regression' of one factor y on another x. Regression analysis is of considerable use in many ecological investigations, but details of the method are beyond the scope of this book and the student may refer to any elementary textbook of statistics. In general a regression analysis is more informative than a correlation coefficient which merely gives information on the level of the relationship. Furthermore if the independent variate x is not normally distributed, the use of the correlation coefficient is, strictly speaking, not valid. It should also be noted that the relationship may in fact not be a linear one and accordingly a correlation coefficient may not reach a significant level. In such a case the calculation of a curved regression (quadratic or high power regression) would be a possible approach.

However, despite these shortcomings, correlation is still of considerable use in ecology. The use and calculation of a correlation coefficient can be suitably demonstrated with the data which have been given in the form of a histogram previously (Fig. 2.7). These illustrate the floristic change along a transect running from a zone of vegetation on clay with flints down into chalk grassland on the North Downs. The actual data are given below in Table 5.1. From a scatter diagram (Fig. 2.4) a clear impression is given of the general negative relationship between the percentage covers of *Agrostis* and *Festuca*.

Table 5.1   (see text)

Agrostis tenuis cover values

x_1

7	11	11	3	23	7	11	10	7	13	21	13	18	8	1
12	9	4	3	3	6	1	9	8	5	4	4	3	8	6
7	5	15	13	21	42	36	40	31	34	19	37	35	47	20
31	36	13	16	22	46	35	31	18	9	27	19	16	22	19

12	21	7	10	9	9	8	–	10	5	–	7	33	9	3	19	15
11	10	25	27	15	7	6	8	14	8	22	17	4	23	7	10	11
23	19	7	7	30	23	16	19	39	35	23	26	32	26	20	21	34
36	27	20	26	29	28	45	43	28	38	25	56	41	28	33	26	37

Festuca rubra cover values

x_2

22	22	33	37	24	22	1	30	19	22	21	19	17	32	43
22	38	25	20	18	34	43	27	13	30	27	15	13	22	15
31	22	15	14	17	10	7	10	24	7	11	4	6	8	8
7	14	27	19	14	5	13	12	27	25	16	8	13	18	–

12	25	26	25	18	16	23	28	32	29	31	28	13	42	28	20	28
15	19	20	8	25	19	27	17	16	22	26	18	23	13	6	14	23
29	13	24	7	2	24	24	21	9	10	20	8	4	12	24	10	6
1	5	1	2	–	7	11	8	12	2	2	1	4	5	–	2	–

The correlation coefficient is calculated as follows:

$n = 128$

$Sx_1 = 2379,$ $Sx_1^2 = 63\ 803,$ $\dfrac{(Sx_1)^2}{n} = 44\ 216$

$$S(x_1 - \bar{x}_1)^2 = Sx_1^2 - \frac{(Sx_1)^2}{n}$$

$$= 63\ 803 - 44\ 216 = 19\ 587$$

$Sx_2 = 2196,$ $Sx_2^2 = 51\ 246,$ $\dfrac{(Sx_2)^2}{n} = 37\ 675$

$$S(x_2 - \bar{x}_2)^2 = Sx_2^2 - \frac{(Sx_2)^2}{n}$$

$$= 51\ 246 - 37\ 675 = 13\ 571$$

$S(x_1 . x_2) = 29\ 939,$ $\dfrac{Sx_1 . Sx_2}{n} = 40\ 815$

$$S[(x_1 - \bar{x}_1) . (x_2 - \bar{x}_2)] = Sx_1 x_2 - \frac{Sx_1 . Sx_2}{n}$$

$$= 29\ 939 - 40\ 815 = -10\ 876$$

Correlation coefficient

$$r = \frac{S[(x_1 - \bar{x}_1)(x_2 - \bar{x}_2)]}{\sqrt{[S(x_1 - \bar{x}_1)^2 . S(x_2 - \bar{x}_2)^2]}} = \frac{-10\,876}{16\,304} = -0.667$$

The number of pairs of observations is 128 and accordingly the table of r-values is entered with 126 degrees of freedom (for table of r-values see Fisher and Yates 1948). For 100 degrees of freedom the probability of exceeding an r-value of 0.3211 by chance is 0.001, i.e. the negative correlation between *Festuca* and *Agrostis* is a highly significant one.

This result is not unexpected and the correlation between the two species is in fact readily appreciated from the scatter diagram (Fig. 2.4). It is often worthwhile to plot a scatter diagram before attempting to calculate a correlation coefficient, since those cases which are clearly not related in any way can be rejected without resorting to what can be a lengthy computation. Valuable information can thus be readily obtained of the level of association between pairs of species or of correlation between a species and the environmental factors, but such analyses can be undertaken only when the initial collection of data is adequately made. When designing a sampling method it is always advisable to bear in mind the possibility that there will be some degree of association or correlation between the species themselves and also with the environment, and wherever possible one should obtain the field data in a quantitative form. Where data exist that are partly (or wholly) qualitative, the analysis can present some difficulty and the resulting information is often limited. (For detailed consideration of such methods the student is referred to Greig-Smith (1957).) *It is important to realize that trends in correlation will depend on the sizes of the quadrats which are used in the sampling* and results should always be considered with this fact in mind.

CAUSALITY OF ASSOCIATION OR CORRELATION BETWEEN SPECIES

Most, if not all, data on the relative abundance of species in a given area will contain one or several correlations of species. The presence of such correlations reflects the pattern or, more usually, several scales of pattern in the vegetation. The existence of pattern in vegetation and the general factors underlying such patterns are dealt with in Chapter 7, but it is necessary to consider at this point in more detail some of the factors that are operative. These causal factors can be studied either by using the techniques available for the detection of pattern (Chapter 7), or by the use of association, correlation, or regression techniques. The two methods of analysis are very closely linked and are only treated separately here for the sake of clarity. The causal factors are split into four sections for convenience, each section represents the level of detail at which it is studied rather than any rigid classification.

SIMILAR OR DISSIMILAR ENVIRONMENTAL REQUIREMENTS

This is probably one of the fundamental reasons for negative or positive relationships that exist between species. On a large scale and with an obvious difference of species composition in two areas such correlations between species lead to the separation of groupings which typify a 'community' (see below and Chapter 9). On a smaller scale such relationships are rarely as obvious and usually need a careful quantitative approach to detect their existence.

An example of such an environmentally controlled relationship is taken from an area of hummocky grassland associated with a calcareous mire. The data for *Carex lepidocarpa*, *C. flacca* and *Potentilla erecta* (Table 5.2) were obtained in the form of frequency, from 150 quadrats positioned by using random co-ordinates (approximately half the sample was related to a hummock). The apparent preference of *Potentilla* for the hummocks and conversely the preference of the *Carex* species for the hollows is readily tested by means of a *t*-test which shows a highly significant difference between the species. This difference is not readily visible on even a close examination of the area though an explanation in general terms is not difficult.

The 'hollows' act as runnels for base-rich water after a period of heavy rain and the pH of the soil is thus neutral to slightly alkaline. Conversely the tops of the hummocks are not affected by calcareous run-off and with leaching over a period of years, together with the steady accumulation of humus, have a pH of about 5·6–5·8 (slightly acidic).

Table 5.2 The relation between topography and the abundance of *Carex lepidocarpa*, *C. flacca* and *Potentilla erecta*

Species	Mean frequency Hummock	Hollow	S.E.	Difference between frequencies	*t*	*p*
Carex lepidocarpa	3·15	5·34	0·4163	2·19	5·260	0·001
C. flacca	0·82	2·0	0·2963	1·18	3·986	0·001
Potentilla erecta	4·19	1·38	0·1967	2·81	14·264	0·001

It is most probable that the distribution of these species is not governed directly by the pH of the soil, but by other factors which are in turn correlated with pH. The relationship between the environment is thus most easily appreciated by relating their abundance in a sample quadrat to pH.

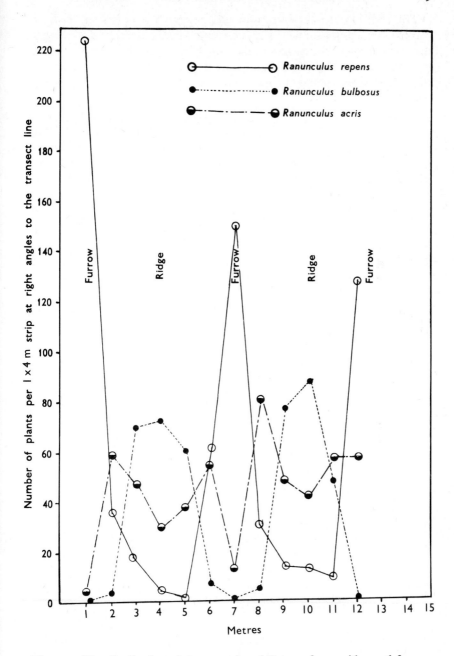

Fig. 5.1. The distribution of three species of *Ranunculus* on ridge and furrow grassland. (From Harper and Sagar 1953; courtesy of British Weed Control Council.)

This relationship of abundance and topography obviously leads to correlations (positive or negative) between pairs of species. In this example the distribution of the species is not sharply controlled by topography but merely influenced differentially by hummocks and hollows. The relationship accordingly cannot be appreciated merely by an inspection of the area.

Harper and Sagar (1953) have shown the spatial distribution of *Ranunculus repens*, *R. bulbosus* and *R. acris* shows a marked correlation with drainage. They investigated the relative abundance of the species along a transect orientated at right angles to furrows and ridges in a permanent pasture. The results (Fig. 5.1) show quite clearly the differential environmental requirements of the three species, *Ranunculus repens* being abundant in the furrows, *R. bulbosus* more abundant on the ridges and *R. acris* showing an intermediate distribution on the sides of the ridges. Seedling establishment under experimental conditions at three levels of drainage suggest it is this part of the life cycle, i.e. germination and early establishment, which is susceptible to the drainage of the area as a whole (Fig. 5.2).

0 Pots with free drainage
½ Pots half water-logged
1 Pots completely water-logged

Fig. 5.2. The early establishment of three species of *Ranunculus* sown in pots with the water table maintained at different levels. Sowings were made on June 26 1953 and counts made on July 28 1953. (From Harper and Sagar 1953; courtesy of British Weed Control Council.)

These results from a markedly variable habitat can be related to the non-random distribution of *Ranunculus* in more homogeneous grassland. Micro-variation of drainage, will modify other nutritional factors to a greater or lesser extent, which will in turn influence the relative abundance of each species in a small area, and lead to negative correlation between them and to marked non-randomness in the population (see also Chapter 7, p. 128).

Both these examples are related to micro-topography and to its effects on abundance of species either directly or, more probably, through other inter-related factors. They illustrate how slight environmental variations can lead to the production of patterns in vegetation and to the existence of correlation or association between species. The choice of examples relating micro-topography and the distribution of species by no means excludes the importance of other variable factors which may operate, but it merely underlines the scarcity of data on such inter-relationships.

MODIFICATION OF THE ENVIRONMENT BY ONE SPECIES ALLOWING ESTABLISHMENT BY OTHER SPECIES

Considerable interest was aroused by the publication of a paper by Went in 1942 which demonstrated a marked relationship between the distribution of annual species and perennial shrub species in a desert area in California. His data showed a general positive relationship between shrubs and annual herbs, but a more striking relationship between annual species and dead *Encelia farinosa* bushes than with living bushes. This he suggested was due to the production of a toxic substance by *Encelia* which accordingly limited the herb population closely adjacent to a living individual bush. Gray and Bonner (1948) subsequently isolated such a substance from the leaves of *Encelia*. Later workers in this field have suggested other equally plausible explanations of the phenomenon in terms of the relative build-up of humus content underneath *Encelia* and other species of shrubs. Whatever the causal explanation (see below for a detailed account of the presence of toxic substances in plants) this work did focus attention on the relationship between an individual and its environment and the possible effects of this interaction on other species within the community.

A clear example of this type of relationship taken from stony desert in Iraq is given by Agnew and Haines (1960). There is a considerable deposition of wind-borne material round the base of perennial plants in this area; this leads to the formation of low mounds around the individual shrubs. *Astragalus spinosus* is an outstanding example of this and there is a marked positive correlation between this species and small annual or perennial herbs, both in the sense of abundance as well as luxuriance. A typical cross-section of such mounds is given in Fig. 5.3. *Astragalus spinosus* is a straggling shrub with tough branches and an abundance of sharp spines which are formed from old leaf axes which have hardened after the leaflets have been shed in the summer. These axes remain firmly attached to the woody stems for several years. The relationship is explained by an interaction of several factors which in turn are related to the

Early mound
formation with
associated herbs

Mature mound with
dense ground vegetation

Fig. 5.3. The marked luxuriance and increased abundance of herbaceous species under *Astragalus*. (From Agnew and Haines 1960; courtesy of *Bull. Col. Sci.*)

growth form and general morphology of *Astragalus*. The first important factor is the protection of the associated herb flora from grazing, which is provided by the spiny branches. This factor is of considerable importance in the Middle East where there is very heavy grazing by goats, sheep and camels even in apparently remote desert areas. Grazing will markedly affect both the luxuriance and abundance of individuals of a species. The second factor is the actual presence of wind-borne material which provides a limited area of soil capable of supporting plant growth. The desert area is largely stony and the soil mounds thus provide an excellent site for establishment and growth of herbaceous species. Further increments of debris are added annually; these include plant remains both from the *Astragalus* as well as from the herbaceous cover, thus steadily raising the humus content of the mounds. This has a two-fold effect; the concentration of soil nitrogen is increased and enables a larger population of herbs to exist, and this in turn reduces wind velocity at the surface of the mound. The reduction of wind velocity lowers the rate of evaporation from the soil surface and in effect leads to a more efficient utilisation of the available rainfall.

Thus it is quite clear that the marked association outlined above is dependent on localized variations of several environmental factors, which in turn are related to the existence of a perennial species which has established itself in the first place. This example is clear cut since the environment is so extreme that it approaches the level below which plants cannot maintain themselves, and a slight (advantageous) shift of the environment produces a marked and a readily visible response in the vegetation. Similar relationships will hold in more temperate regions of the world, but here they are much less obvious and little is known about them at present.

The fact that vegetation modifies and changes the environment to a considerable extent is largely responsible for plant succession in general (see Chapter 3). Numerous examples of environmental modification by vegetation are available (e.g. Fig. 3.6, p. 47) and quantitative data are

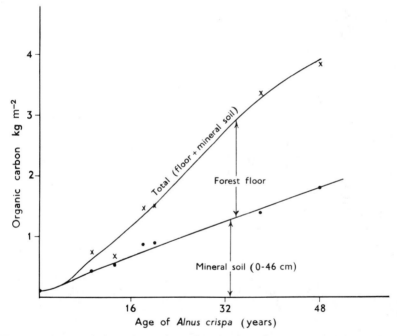

Fig. 5.4. Accumulation of organic carbon in the mineral soil and forest floor under *Alnus crispa* in Alaska. (From Crocker and Major 1955; courtesy of *J. Ecol.*)

given by Crocker and Major (1955) relating the accumulation of organic carbon to the age of a stand of *Alnus crispa* (Fig. 5.4). This accumulation of carbon in a 50-year period and the resultant changes in the vegetation is in fact a characteristic and well marked succession (see Chapter 3, p. 40).

Moreover, this example illustrates the steady build up of humus material in relation to vegetational cover. The rate of build up of humus is of course also dependent to a considerable extent on climate and drainage. Thus in *Astragalus* the nitrogen accumulation will be relatively small, but sufficient to have an effect on the associated vegetation, which leads to the marked positive correlation of species. It is a reasonable assumption that similar effects are present in temperate herbaceous vegetation, but the relative infrequent occurrence of strong positive association between species in this type of vegetation would suggest that such modifications of the environment are often overshadowed by environmental variations resulting from other factors.

PRODUCTION OF TOXIC SUBSTANCES BY PLANTS

As far back as 1832 De Candolle suggested the importance of root excretions and since 1900 considerable attention has been directed towards the detection of substances in plants which are toxic to other species. The demonstration of direct chemical interaction between species is surprisingly difficult, and much of the earlier work designed to prove an interaction between two species based on the exudation of a toxic substance can be criticized on the grounds of experimental design or may be explained in terms of competition for some environmental factor.

Bedford and Pickering (1919) examined in some detail the vigour of apple trees in relation to the grass cover of an orchard. Observation showed a very marked difference between the height and fruit production of trees grown in tilled soil compared with trees of similar age grown in plots with a complete grass cover. For example a plantation of Bramley's Seedling apples established for 23 years was selected and half of it laid down to grass. In the first year the grass did not establish well but even then the season's crop was reduced by 5 per cent (compared with the crop from the tilled section). In the next year the crop was markedly affected as fruit production was reduced by 89 per cent following the complete establishment of the grass. Similar results were obtained in several other experiments, all of which strongly suggested the action of a toxic substance produced by the grass. This possibility was examined further experimentally. Young apple trees were grown in pots while a 'surface crop' of grass was suspended above in a perforated container. Water was supplied through the surface crop which leached any toxic substance into the pot below. The resultant inhibition by the surface crop was very marked in most experiments and amounted to almost complete suppression of tobacco plants by clover and to 40–60 per cent suppression in other experiments.

The results at first sight seem conclusive but the field observations can be explained in part by competition for nitrogen and water. The inhibition of the trees by grass cover can be almost completely eliminated by addition of nitrogenous fertilizers, together with irrigation of the orchard. The pot experiments are quite convincing, however, and one is forced to conclude that the presence of a toxic substance is probable, but under field conditions it is either destroyed by bacterial action or is absorbed by soil particles. It should, however, be borne in mind that the removal of an inhibitory effect by the addition of nitrogen and water does not necessarily prove that a toxic substance is *not* operative. It could be that the toxic substance produced by the grass for example reduces the root development of the apple trees and accordingly limits their ability to take up nitrogen and water. With the availability of abundant nitrogen and water the limiting effect of a reduced root system is minimal and the apparent toxic effect disappears. In other words the mechanism of competition for nitrogen and water could be through the action of the toxin.

The relationship between apple trees and surface crop illustrates clearly the variety of suggestions that can be offered as an explanation of an observed result. The experiments have to be designed with considerable care, and the above example does emphasize the attendant difficulties in this field of research. More recent work has revealed numerous examples of inhibition of one species by another; these have been usefully reviewed by Bonner (1950) and Woods (1960), and several well documented cases exist of marked inhibition of one species by another where the toxic substance presumably responsible has been isolated and identified. Thus it has been shown that Scopoletin is liberated from oat roots (Martin 1956, 1957, 1958) and that it possesses growth inhibiting properties (Goodwin and Taves 1950; Libbert and Lübke 1957, 1958). Similarly Gray and Bonner (1948) isolated 3-acetyl-6-methoxy-benzaldehyde from leaves of *Encelia farinosa* which was shown to be toxic to tomato and corn seedlings, but it was noted that the toxic effects were far less effective in good garden soil than in sand culture. This suggests an inactivation or destruction of the compound by the soil microflora. The most impressive example is that of *Juglans* (walnut) which has a marked effect on some associated species (Jones and Morse 1903). Under field conditions it will produce wilting of potato and tomato plants grown in the immediate vicinity (Cook 1921), and has an identified toxic constituent, juglone, present in its roots (Davis 1928). Purified extracts of juglone have produced toxic symptoms in a variety of species investigated so far (Massey 1925, Perry 1932).

It is clear that in many cases toxic substances are produced which will inhibit other species under experimental conditions. However, it is equally clear that usually the activity of the toxin is markedly reduced in soil, probably by the action of soil micro-organisms. Also it is questionable whether sufficient quantities of toxin capable of exerting a measurable effect on associated species are released in the soil. Thus in terms of toxic substances no conclusive proof exists which would adequately explain positive or (more usually) negative correlations between species. However, the possibility that such a mechanism may operate in some cases at least should not be ignored and the case of *Juglans* (see above) is probably the best available example to date. As has been suggested above, there is considerable difficulty in differentiating under field conditions between a toxic effect and one due to competition. This situation is made even more complex by the possibility that the different competitive ability is actually achieved through the agency of the toxic excretion.

COMPETITION LEADING TO NEGATIVE AND POSITIVE CORRELATION OR ASSOCIATION

The term 'competition' is used in a wide variety of contexts with a slightly different meaning and it has been suggested that 'interference' (Harper 1961) be used as a general term to describe the loss of vigour and productivity of an individual due to the close proximity of another. In this context, however, the term is used to describe the struggle between individuals for some environmental factor such as light or nitrogen, and the final effect is not one of complete elimination of one individual by another but a lowering of the performance of one of the competing members. This is usually manifested in a reduction in the numbers of a plant in relation to another species and leads to a negative correlation between the two.

Much of the detailed work on competition has been carried out under controlled conditions, since in the field it is often difficult to distinguish factors involved in competition from the other possible causes of a negative correlation. In a consideration of effects of competition it is important to realize at which point of the life cycle competition exerts the most pressure, whether it is at the seedling stage or whether the effect is over a longer period of time. The effect of marked competition at seedling stages is usually of greater importance in relation to the actual numbers of plants which survive to maturity. Reduction in numbers of plants is more easily observed in ecological situations than the less obvious effect of reduced performance caused by the close proximity of a competitor. Conversely, in an agricultural or horticultural field, performance and yield are of paramount importance and are related both to plant numbers as well as to the general level of performance of the crop.

An example of the competition between individuals of the same species is given by Harper (1960) who investigated the numbers of mature plants of four species of poppy in sample plots; the numbers resulted from different densities of sown seed. The results (Fig. 5.5) show a clear decline in the number of mature plants relative to the rise in the density of seeds per plot. The density of sown seed produces a corresponding reduction in

Fig. 5.5. Competition between individuals of the same species, as expressed by number of mature plants per square foot (*a*), and capsules per square foot (*b*), in plots sown with different densities of seed. (From Harper 1960; courtesy of Blackwell Sci. Pub.)

the number of capsules per plant produced by the mature plants. Similarly Donald (1951) (see also Harper 1961) gives data relating the plant weight of *Trifolium subterraneum* to the density of sowing at successive dates (Fig. 5.6). At the time of sowing individual plant weight is of course not dependent on density, but as the plants develop those at the highest density compete for available nutrients and light and show a reduction of plant weight. This effect is seen earliest at high densities but is apparent at progressively lower densities as time goes on and the plants grow larger. These examples show quite conclusively the dependence of plant numbers

Fig. 5.6. Reduction of individual plant weight by competition between individuals of *Trifolium subterraneum* at different densities and at different times from sowing. (From Harper 1961; courtesy of Cambridge Univ. Press.)

and their performance on initial density of seeds and on the density of mature individuals respectively.

Data showing the effect of different species on the survival of *Juncus effusus* seedlings at different levels of soil fertility are given by Lazenby (1955). Five combinations of species were used: *Juncus* growing alone (J), *Juncus, Molinia caerulea*, and *Trifolium repens* (JTM), *Juncus* and *Trifolium* (JT), *Juncus, Trifolium* and *Lolium perenne* (JTL) and *Juncus, Trifolium* and *Agrostis tenuis* (JTA). The results are summarized in Fig. 5.7 and all show a marked reduction of both the numbers of plants as well as the performance of *Juncus* in all the competitor combinations used, and at all nutritional levels. Lazenby suggests that the important effects are on the early seedling states of *Juncus* since its seeds require light before they will germinate. Accordingly those species which form a very close sward inhibit germination to a considerable extent.

The examples discussed above are all taken from experiments performed under fairly well controlled environmental conditions, but they all demonstrate the sort of effect which can be expected in natural vegetation. Well documented examples from natural vegetation are few and perhaps

Fig. 5.7. Relative plant numbers, tiller numbers, weight per tiller and total weight of *Juncus effusus* developed with different companion species during 18 months' growth. (From Lazenby 1955; courtesy of *J. Ecol.*)

the most informative data are those relating to the interactions between a tree canopy and the ground vegetation. It is sometimes assumed that the paucity of vegetation beneath trees is a direct result of the low light intensity that prevails at ground level for most of the growing season. This is by no means always correct. Watt and Fraser (1933) examined this relationship in a *Pinus sylvestris* woodland with a ground flora dominated by *Deschampsia flexuosa* and *Oxalis acetosella*, and related the abundance and vigour of these two species to competition with the tree roots for some nutrient factor, possibly nitrogen. The experimental layout consisted of a series of plots surrounded by trenches, which severed tree roots and isolated the herb layer from competition with tree roots at different depths. Five trench depths were used 2·5–5 cm (raw humus), 8–10 cm (raw humus cut to its entire depth), raw humus plus 10 cm of mineral soil, raw humus plus 20 cm of mineral soil, and raw humus plus 33–45 cm of mineral soil. In addition a control plot, and a series of plots which were not trenched but to which distilled water or nitrogen were added, were also sampled. The plots were sampled prior to trenching and for two years subsequently, after which the experiment was discontinued. The samples were oven dried and the results expressed as percentages of the original crop in each plot. The maximum rooting depths of *Deschampsia* and *Oxalis* were observed to be approximately 60 cm and 8 cm respectively. The majority of the *Pinus* roots were confined to a depth of 8–20 cm. A similar series of experiments were also conducted under beech canopy in an adjacent wood.

The results are summarized in Fig. 5.8 below. In assessing these results note must be taken of the annual reduction of the plot herbage caused by

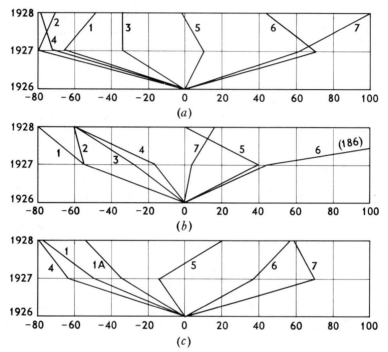

Fig. 5.8. Percentage increase or decrease in the weights of the original field layer of *Deschampsia flexuosa* under pine (*a*), *Oxalis acetosella* under pine (*b*), and *O. acetosella* under beech (*c*). The treatments were: 1. Control. 2. Distilled water added. 3. Raw humus cut 2·5–5 cm. 4. Raw humus cut to surface mineral soil (8–10 cm). 5. Raw humus plus 10 cm mineral soil cut. 6. Raw humus plus 20 cm mineral soil cut. 7. Raw humus plus 33–45 cm mineral soil cut. (From Watt and Frazer 1933; courtesy of *J. Ecol.*)

the actual harvesting. However, they do allow several preliminary deductions to be made. Firstly, the addition of water had no effect on either *Deschampsia* or *Oxalis* (Fig. 5.8, *a*2 and *b*2). Secondly, that cutting through the humus and 10 cm of mineral soil and through the main roots of the trees enables both herb species to hold their own and probably in the absence of harvesting they would have actually benefited. Trenching to greater depths shows a general increase in crop weight (Fig. 5.8 (6) and (7)). The addition of nitrogenous fertilisers gave mixed results, *Oxalis* showed no change when the rates of application were small but was completely eliminated by the larger dosage. However, *Deschampsia* benefited from the small doses although the largest dose had an adverse effect. Thus it appears that nitrogen may be one of the important factors for *Deschampsia* at least, which is limited by tree root competition. Other possible effects of the trenching which should also be noted are that the subsequent death and decay of the severed tree roots would increase the level of nutrient in the soil to a slight extent and also that the decay of the roots would allow greater rates of oxygen diffusion along 'channels' once filled by the roots. The increase in vigour of the herbaceous vegetation is sufficiently large to suggest an increased benefit which is greater than these two possibilities would provide. It is remarkable that the most marked effect on both herb species is at the maximum depth of trenching even though *Oxalis*, for example, has a shallow rooting system.

The inhibition of a stand of trees by the ground vegetation is perhaps, at first sight, surprising, but is a well-known factor in the establishment of certain tree crops and offers a further example of what has been interpreted as competition. One of the more valuable crops planted by the Forestry Commission is Sitka spruce (*Picea sitchensis*), as its yield of timber is high. Early trial plantings showed that spruce planted in *Calluna*, though establishing fairly well, was immediately 'checked' in its growth. Addition of basic slag, as a source of nitrogen and phosphorus improved growth initially, but did not prevent the final onset of 'checking'. The experimental plantings included some mixed stands of Sitka spruce and Scots pine and it was noted that the growth of spruce was greatly assisted by the presence of pine. Weatherell (1957) gives data comparing two stands of equal age (Table 5.3) which show quite clearly the increased growth after 10 years which is even greater after a longer period of time (Table 5.4). This use of a 'nurse' crop is now standard practice. Japanese larch is more commonly used than pine, and the apparent beneficial effect of one tree species on another is related to the presence of *Calluna vulgaris* as a dominant species of the ground layer. Observations

Table 5.3 Comparison of Sitka spruce pure and in mixture with Scots pine at 10 years of age. (From Weatherell 1957; courtesy of *Quat. J. Forestry*)

	Pure plantation (Sitka spruce)		Mixed plantation (Sitka spruce and Scots pine)	
	Mean height spruce (feet)	Mean growth 1937 (inches)	Mean height spruce (feet)	Mean growth 1937 (inches)
Not manured	1·8	1·4	2·4	4·1
Basic slag applied shortly after planting	2·6	3·4	3·2	6·6

Table 5.4 Comparison of Sitka spruce pure and in mixture with Scots pine at 12 years of age. (From Weatherell 1957; courtesy of *Quat. J. Forestry*)

	Pure plantation spruce		Mixed plantation spruce and pine	
	Mean height (feet)	Mean growth 1939 (inches)	Mean height (feet)	Mean growth 1939 (inches)
Not manured	2·0	1·7	3·2	6·4
Basic slag applied after planting	3·1	2·8	4·2	7·0

show that the roots of spruce growing adjacent to Japanese larch tend to be very much more developed under the larch than in the area dominated by *Calluna* and tend to run in the larch needle litter, only rarely descending into the mineral soil. Japanese larch very effectively suppresses *Calluna* and concurrently with the removal of this competition spruce will grow at a greater rate than a pure stand which only slowly suppresses *Calluna*. It seems likely that the roots of larch and spruce occupy different levels in the soil and accordingly any competition between them is minimized. The effect of Japanese larch as a nurse crop is well marked (Table 5.5), the height of spruce being doubled over a period of 13 years. The beneficial effects of basic slag suggest that nitrogen (the addition of

Table 5.5 Height in feet of various species mixtures after 13 years. (From Weatherell 1957; courtesy of *Quat. J. Forestry*)

Crop	Height of spruce (feet)	
	Complete ploughing	Single furrow ploughing
Three rows of spruce alternating with two rows of Japanese larch	13·2	12·4
Sitka spruce plus broom	13·3	9·0
Two rows of spruce alternating with two rows Scots pine	9·4	8·5
Alternating rows of spruce and pine	10·5	10·6
Two rows of spruce alternating with single row Scots pine	7·5	8·5
Pure stand spruce	6·0	6·0

calcium accelerating nitrification) or phosphorus or both are possible requisites in the competitive interaction between spruce and *Calluna*; however, other factors may well be involved.

The relationship between the ground layer and tree layer outlined above demonstrates a presumed competition between the two life forms. None of the results is conclusive and the same difficulty exists in attempting to relate the observed effect with a causal mechanism. That is to say, it seems a reasonable assumption that the observed increase in vigour in these instances is directly related to the removal of competition, but it can be argued that an identical effect would be produced by the removal of the source of an inhibitory substance. This is the converse of the problem outlined above (p. 95); when the results of controlled experiments are extrapolated into field conditions the definition of causality is immediately open to several explanations unless the data is obtained by extremely critical sampling of both vegetation and environment. Due to the time factor, such sampling presents considerable difficulty, even if one assumes that soil analyses can be done to the required order of accuracy; this latter difficulty at the moment appears insurmountable.

6 Plant Population Dynamics

STUDIES of animal populations involving a census of their numbers have occupied a central position in animal ecology for a considerable period of time. There has, however, been no equivalent work on population numbers of plants. Plant ecology has been largely dominated by investigations into the spatial configurations of species in relation to their environment. At a small scale such configurations are termed pattern (Chapter 7) and at larger scales, plant associations, which can be characterized by their 'unique' species spectrum (Chapters 9 and 10). Until relatively recently very little attention has been focused on plant populations and the control of the numbers of individuals involved. This lack of information partly stems from the pre-occupation with spatial configurations and partly from the difficulties inherent in studying changes in natural populations of plants: In animal populations changes can be fairly readily documented by changes in numbers. In plant populations, changes in numbers of individuals do occur, but more frequently a population change is initially reflected by much less obvious criteria. Changes in dry weight, growth morphology, number of seeds per individual, all indicate pressure on the population from environmental changes or by increased interference from other species. A change in actual numbers of mature individuals is a much more gradual process, especially where perennial species are involved. Understandably what information is available is largely derived from experimental studies on populations of annual weed species, which with their short life cycle are more amenable to an experimental approach. The limited evidence available for perennial species suggests that at least the numbers of the population established by seed are controlled in the same way as annual species.

The main factors that so far have been shown to affect population numbers are summarized as a simple flow diagram for two hypothetical species A and B (Fig. 6.1). Each stage of the life cycle which interacts with or is affected by either an environmental parameter or another species is shown. Equally the overall importance of season, biotic factors, and genetical make-up of the potential population is indicated:

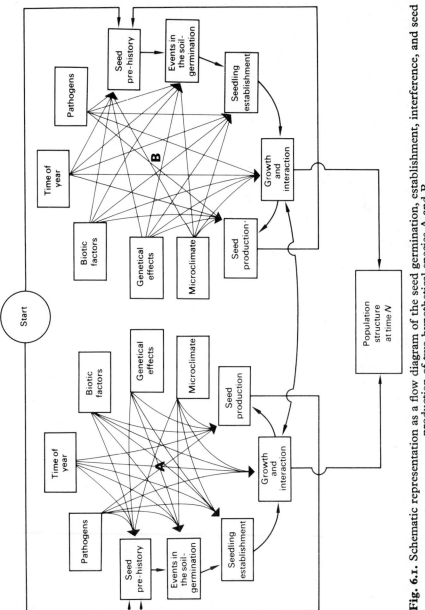

Fig. 6.1. Schematic representation as a flow diagram of the seed germination, establishment, interference, and seed production of two hypothetical species A and B.

1. EVENTS IN THE SOIL

The numbers of individuals reaching maturity in a population at any given point in time is clearly not only a function of how many seeds are available, but also of what physiological state they are in. Many plant species produce seeds which do not germinate immediately but which lie dormant until a particular sequence of either intrinsic events, extrinsic events, or both, trigger germination. Harper (1957) conveniently classifies dormancy into three categories: Innate dormancy, which is a specific property of a species and requires a particular treatment to break it, induced dormancy which can develop when a non-dormant seed is exposed to a particular (harsh) environmental situation, and enforced dormancy developed in response to external conditions but removed immediately when the conditions are suitably modified.

Increasing O_2 tension around the seed embryo by pricking or abrasing the seed coat, giving a fluctuating temperature or low temperature treatment, light treatment, or an increased level of nitrate availability, have all been shown to induce germination [Stokes (1965), Thurston (1959), Toole et al. (1956), Vegis (1963)]. All of these parameters are likely to be operative in all plant systems. Seed coats decay with time allowing an increased oxygen supply to the embryo and equally temperature fluctuates seasonally. Since nitrate concentration also varies throughout the season due to the temperature dependence of nitrogen-fixing organisms (Russel 1961), the control of the actual number of seeds available for germination at time N is extremely complex. Popay and Roberts (1970a and b) examined the emergence of populations of the weed species Senecio vulgaris and Capsella bursa-pastoris over a two-year period and correlated in part the observed flushes of germination with the breaking of dormancy.

The germination requirements of Senecio are a light treatment (which is essential for germination), and a temperature of 10–15°C (Fig. 6.2).

Fig. 6.2. Effect of different constant temperatures on the germination of seeds of (a) Capsella and (b) Senecio when exposed to light. ▲, after two days in Senecio or after five days in Capsella; ○ after 10 days, ●, after 14 days; ×, after 28 days. (From Popay and Roberts 1970; courtesy of J. Ecol.)

In addition dry storage at temperatures up to 35°C further increases germination (Fig. 6.3). Complete suppression of germination is produced by darkness, low temperatures, low oxygen supply and high CO_2 concentration. Conversely *Capsella* requires as an essential factor a low-temperature treatment (below 10°C) as well as light. Dry storage has

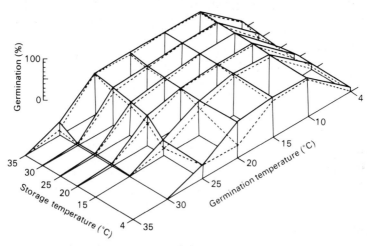

Fig. 6.3. The effect of dry storage at different temperatures on the germination of *Senecio* seeds over a range of temperatures. ——, Germination after 8 weeks' storage; – – –, after 4 weeks' storage. The seeds were exposed to diffuse light for a few minutes each day. (From Popay and Roberts 1970; courtesy of *J. Ecol.*)

little or no effect. Popay and Roberts also showed a marked interaction between per cent germination and concentration of nitrate whilst a fluctuating temperature régime was employed in the experimental lay-out. Low concentrations of nitrate solution (10^{-1} or 10^{-2} mol dm^{-3}) coupled with a diurnal temperature fluctuation of 19 h at 4°C and 5 h at 30°C gave a very marked response (Fig. 6.4).

The resultant ecological implications are clear-cut: The germination behaviour of freshly collected *Senecio* seed varies tremendously throughout the year (Fig. 6.5). There is little dormancy in seeds collected in the summer, whilst conversely, material collected in the spring showed only 30 per cent germination at the optimum temperature of 10°C. Thus the dry storage treatment (see above) seems to be effectively accomplished during July and early August, whilst the seeds are ripening on the plant, allowing subsequently a wide temperature range for successful germination. Their results indicate a burst of germination in *Senecio* as the soil temperature rises above 10°C in January and again in February (Fig. 6.6). There is also a marked emergence of a further population of both species

Fig. 6.4. Germination of *Capsella* seeds in the presence of inorganic nitrogen sources at different concentrations and temperatures. 4/30°C refers to diurnal fluctuations in the ratio of 19/5h between 4° and 30°C. The tests were continued for 28 days, during which the seeds were exposed to light for a few minutes each day. (a) 10^{-1} M solution. (From Popay and Roberts 1970; courtesy of *J. Ecol.*)

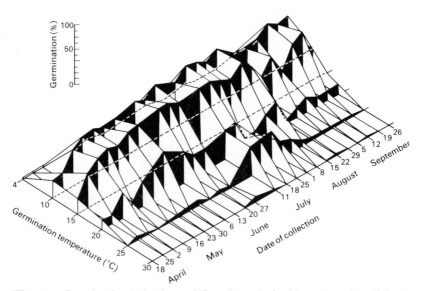

Fig. 6.5. Germination behaviour of *Senecio* seeds freshly collected at different times of the year from the same habitat. The seeds were exposed to diffuse light for a few minutes each day during the 32-day germination tests. (From Popay and Roberts 1970; courtesy of *J. Ecol.*)

in June to August which is correlated with maximal diurnal soil temperature fluctuation and simultaneously an increase in soil nitrate level.

Other sources of variation in germination of seed populations are known. Black (1955) has shown that depth of sowing affects the pre-emergence weight changes in *Trifolium subterraneum* and particularly cotyledon area, which has considerable effects on the subsequent performance of the plant (see below). This parameter is only likely to affect those plant populations where there is a regular soil disturbance factor, notably populations

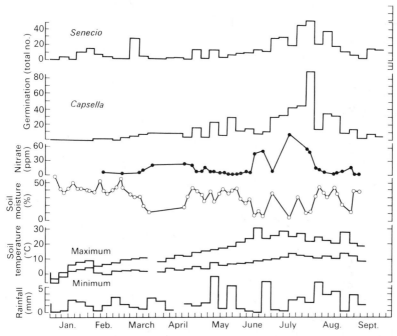

Fig. 6.6. Weekly emergence of seedlings at the experimental site in 1967. The values for the experiments on natural seed populations are the total numbers of seedlings emerging each week. Nitrate concentration is expressed in parts per million (p.p.m.) of nitrate in solution in the soil water. (From Popay and Roberts 1970; courtesy of *J. Ecol.*)

of weed species. The possibility of a biotic interaction and the burying of seeds by such a vector cannot be ignored, however, in natural populations. Watt (1919) showed, for example, that the viability of acorns depended on their per cent moisture content being maintained by trampling into the soil by grazing animals. Other similar relationships may also exist.

The size of a seed can have considerable control over its subsequent germination pattern and growth. Black (1956) using subterranean clover and three seed size classes of mean weights 3 mg (or less), 5 mg and 8 mg respectively, showed a marked relationship with both dry weight of seedlings and establishing plants as well as leaf area (Fig. 6.7). Similarly, Cavers and Harper (1966) show slight but significant differences in seed weight taken from panicles on the same plant but from different parts of the same panicle, as well as at different harvest times from the same individual plant. These differences are correlated with percentage germination data obtained under alternating temperature (15–30°C) in the dark with a measure of seed viability using a light treatment with an alternating

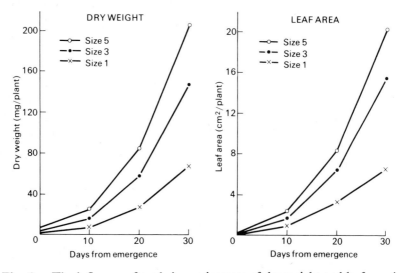

Fig. 6.7. The influence of seed size on increase of dry weight and leaf area in subterranean clover. (From Black 1956; courtesy of *Aust. J. agric. Res.*)

temperature of 10–20°C giving 100 per cent germination. The seeds all show a uniformly high level of viability yet with clear differences in germination response to the test conditions, in addition to the marked differences between the replicates A-D collected from separate locations (see below).

The implications of these results on population size and population dynamics are of obvious importance. Clearly the balance between species composition will depend in part on the climatic régime of the preceeding months interacting with the particular dormancy/germination characteristics of the species components. Different sites having slightly different microclimates will as a result carry slightly different species populations. For perennial species, originally having established by seed, these differences could be perpetuated or even intensified by subsequent interactions (see below).

Harper *et al.* (1965) have further demonstrated the importance of both soil surface texture, and small modifications of soil microtopography on germination of *Plantago* and *Bromus* species. Sheets of glass were used to make square depressions 1·25 cm and 2·5 cm deep, or left on the soil surface. Sheets of plate-glass were inserted vertically into the soil leaving different heights projecting above the soil surface and aligned either N-S or E-W. Thin-walled wooden boxes without top or bottom were also inserted to different depths in the soil. The results (Fig. 6.8) show a

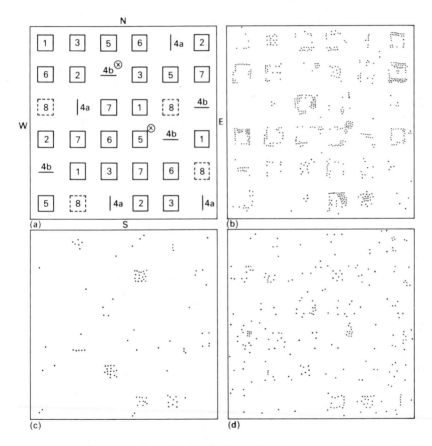

Fig. 6.8. (a) Plan showing the distribution of various types of objects (1–8) placed on a soil surface sown with seeds of *Plantago* species. ⊗ = worm casts. (b) The distribution of seedlings of *P. lanceolata* in relation to objects and depressions on the soil surface. (c) As for (b) with seedlings of *P. media.* (d) Seedlings of *P. major.* (From Harper *et al.* 1965; courtesy of *J. Ecol.*)

marked positive correlation of position of germination with the different objects and surface microtopography. Additional experiments using soil clods of different grades, as a surface layer on top of a soil–sand mixture, again demonstrated a marked relationship between surface irregularity (expressed as microtopographical variance) and numbers of seeds germinating (Fig. 6.9). Furthermore, *Bromus madritensis* established much

Fig. 6.9. The relationship between the numbers of seeds of *Bromus* species sown and the numbers of seedlings established on soils of varied microtopography. (□) Soil microtopographical variance = 189·5 mm^2; (○) = 166·5 mm^2; (■) = 60·6 mm^2; (●) = 2·8 mm^2. (From Harper *et al.* 1965; courtesy of *J. Ecol.*)

more effectively on rough-surfaced soils than *B. rigidus*. Seedlings of Kale and *Chenopodium album* showed a similar response to surface roughness establishing much better on rough surfaces than on smooth. These roughness effects could be cancelled out by covering with polythene covers, and Harper *et al.* interpret this result as indicating seed–water relationships to be the factor affected by surface roughness, with temperature playing a lesser part. However, in the light of the findings of Popay and Roberts (1970*a* and *b*) (see above) the relationship may be much more complex.

In addition to the control of seed dormancy and germination by the environment, there exists a considerable number of soil pathogens capable of reducing the seed population prior to germination (Wernham 1951). Mortality of seeds prior to germination and emergence, as a result of exposure to unexpected low temperatures, is a further source of loss of viable seeds. Harper *et al.* (1955) show clearly that maize grains (4 varieties used) are unable to tolerate the cold and wet soil conditions of late autumn through to spring in Britain. This factor alone, over and above any requirement of a longer growing season, limits the extension of maize into temperate regions. Similar considerations will apply to other 'exotic' species and raises the opposite problem of how in fact the seeds of indigenous species do successfully overwinter. Clearly other factors are involved in the soil about which we have as yet, no detailed information.

2. ESTABLISHMENT AND GROWTH

Harper and McNaughton (1962) showed that the number of mature plants of *Papaver* species, sown in pure and mixed populations does not relate to the number of seeds sown, above a certain density of seed. Self-thinning occurs, minimizing the interference between individuals (see also Chapter 5, pp. 85–103), and furthermore, 'The density-induced establishment risk for each species in a mixture is largely controlled by its own density.' In addition to changes in numbers of individuals reaching maturity, the dry weight of mature individuals is also profoundly affected [see also Harper and Chancellor (1959), and Aspinall and Milthorpe (1959)]. Yoda *et al.* (1963) made a survey of a number of species and proposed a relationship $W = C.p^{-3/2}$ between dry weight of the surviving plants and density, where W is the dry weight, p is the density of surviving individuals, and C is a constant related to the growth architecture of the species.

Thus if a plant has a linear dimension L, covering an area S then:

$$S \propto L^2 \tag{1}$$

and
$$W \propto L^3 \tag{2}$$

From 1 and 2
$$S \propto W^{2/3} \tag{3}$$

Since density is also related to area

$$S \propto \frac{1}{p} \tag{4}$$

From 3 and 4
$$W^{2/3} \propto \frac{1}{p} \tag{5}$$

$$W^{-2/3} \propto p$$

$$W \propto p^{-3/2}$$

$$W = C.p^{-3/2} \tag{6}$$

The full significance of this 3/2 power law was recognized by White and Harper (1970) who extended its application to a wide range of both species and species mixtures (Fig. 6.10), and showed it to be of general

Fig. 6.10. Changes in numbers and individual plant weight with time of *Brassica napus* (———) and *Raphanus sativus* (. . . . .) in pure stands and in mixtures; the mixtures consisted of 37 per cent *Brassica*, 63 per cent *Raphanus* (–·–·–) and 66 per cent *Brassica*, 34 per cent *Raphanus* (– – –) 2 weeks after sowing, before changes in numbers occurred. Three fertility levels (○, low; □, medium; △, high) were imposed on the populations. Harvests were taken at $6\frac{1}{2}$ (○), 13 (◉) and 17 (●) weeks after sowing. Confidence limits are drawn at $P = 0.05$. (From White and Harper 1970; courtesy of *J. Ecol.*)

application. Root and shoot components of plant yield appear to follow the same law and equally the yield tables employed by foresters also closely coincide. In addition they tested and confirmed the assumption that it is the smallest plants in a population which are the first to thin, leaving the larger ones to grow more rapidly. This is an important relationship and enables the subsequent pathway of a population to be predicted from initial events. The cause of the thinning phenomenon is not clear. However, the factors influencing seed germination discussed above indicate some of the likely parameters and emphasize the irregularity of emergence likely from a population of seeds. There will be an immediate disparity in age imparting morphological differences which will be rapidly intensified by effects of seed size and soil microtopography. In turn this will influence and alter the canopy nannoclimate generating still further differentials within the population over a period of time.

3. GROWTH AND INTERACTION

Black (1955, 1956, 1958 and 1960) in an elegant series of papers examined the population numbers of subterranean clover, and how they changed. They are especially interesting since they demonstrate one of the mechanisms of interference pressure between individuals.

The influence of seed size on the area of the cotyledons has been discussed briefly before (p. 110, Fig. 6.7) and similarly the interaction with depth of sowing. Black (1955, 1956) showed that seed size determined the depth from which successful seedling emergence can take place and also the initial area of cotyledons. It appears that once successful emergence has been achieved the area of the cotyledons determines the extent of the difference in the early stages of growth. It should be noted that this relationship only holds for seeds with epigeal germination and where the whole seed is embryo plus coat and there is no endosperm.

These early differences are maintained during subsequent development not only in the first weeks (Fig. 6.7 above) but also during establishment of a sward when there is any interference pressure. Not only do numbers of individuals in the population alter at differential rates, but also dry weight increases per plant, dry weight increase per area, leaf area per plant and leaf area per unit of ground (Figs. 6.11 and 6.12). Black (1958) analysed the distribution of foliage in relation to height above ground (Fig. 6.13). The leaf area in a particular layer is represented by the length of a horizontal line at the appropriate height on the y-axis. Pure swards grown from two different seed sizes show closely similar results, both producing the bulk of their leaves between 25 and 30 cm above ground and by the final harvest (August 16) also having little difference in dry weight yield or leaf area (Figs. 6.11 and 6.12). The sward from mixed seed sizes is markedly different. The leaves of the small-seeded plants are shown to the left and the large-seeded plants to the right of the vertical axis. The spatial distribution of the leaves is quite different, the leaves of the large-seeded plants completely dominating the small-

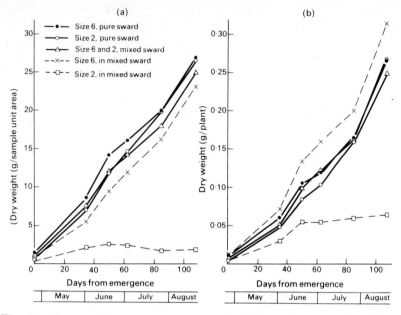

Fig. 6.11. Dry weight increases, per square link (a), and per plant (b). (From Black 1958; courtesy of *Aust. J. agric. Res.*)

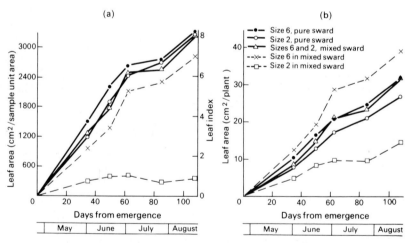

Fig. 6.12. Increases in leaf area, per square link (a), and per plant (b). (From Black 1958; courtesy of *Aust. J. agric. Res.*)

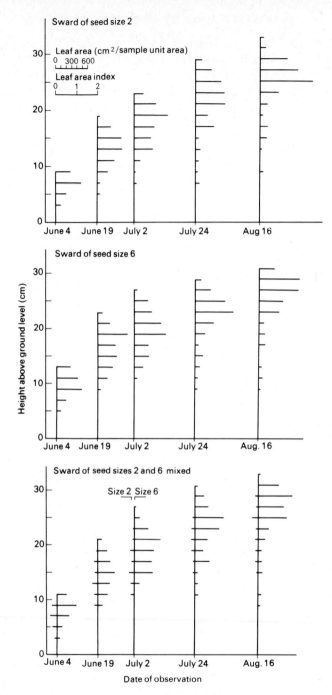

Fig. 6.13. Leaf area at each 2-cm layer of large-seeded, small-seeded, and mixed swards. (From Black 1958; courtesy of *Aust. J. agric. Res.*)

seeded plants. The light microclimate of the sward (Fig. 6.14) shows typical light profile curves, the only marked differences being the relative position of each curve at each harvest date. The initial effect of cotyledon area can be seen in the light profiles for June and early July where the small-seeded plants are shorter, this effect being eliminated by August 16.

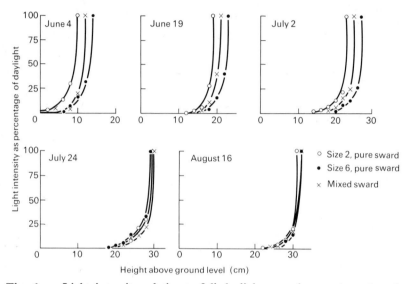

Fig. 6.14. Light intensity relative to full daylight at each 2-cm layer in subterranean clover. (From Black 1958; courtesy of *Aust. J. agric. Res.*)

The plants from small seeds are thus at a continual disadvantage *when grown in mixed populations*: They emerge with smaller cotyledons which limits their early growth thus allowing the rapid dominance of large-seeded individuals. This in turn alters the light microclimate and further checks the development of the small-seeded individuals. In pure swards, however, small-seeded individuals behave in an identical fashion to large-seeded individuals after they have overcome their initial slow start. It is significant that the change of population number is at the expense of small-seeded individuals only, in mixed swards (Fig. 6.15).

An interesting comparison with Black's findings is given by Harper and Clatworthy (1963), using *Trifolium repens* and *T. fragiferum* which have markedly different seed sizes (Fig. 6.16). Growth of pure swards shows that *T. repens* develops more rapidly in the first 14 weeks than *T. fragiferum* and has a higher leaf area index (the ratio of leaf area per unit of ground covered—L.A.I.) (Fig. 6.17); subsequently, *T. fragiferum* develops a higher L.A.I. which levels out around 16–18 weeks, whilst that of *T. repens* falls off reflecting its growth habit (the results were partly affected by rotting of some leaves after lodging). In mixed swards, a similar sequence of events occurs with *T. repens* maintaining a slightly higher L.A.I. and both species coming into equilibrium with each other at 15 weeks. Similarly the vertical distribution of leaf area in relation to

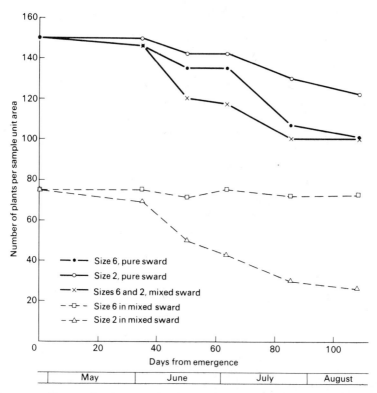

Fig. 6.15. Changes in plant numbers for large-seeded, small-seeded, and mixed swards, and for both components of the mixed sward. (From Black 1958; courtesy of *Aust. J. agric. Res.*)

light profiles indicates a sharing of the incoming energy, substantiating the trend of a satisfactory level of cohabitation of the two species. The results are particularly interesting in that they relate to the establishment of perennial species.

Thus seed size is merely one factor in a complex system which interacts with one or several other factors to decide the final outcome of population balance. Black's data demonstrates a situation where the initial slight advantage of one seedling over another becomes exaggerated by other factors in the system. Conversely, Harper's data shows a system where the reverse is true and the potential deleterious effects of seed size are eliminated by other factors in the system, again emphasizing the extremely complex control mechanisms of population numbers.

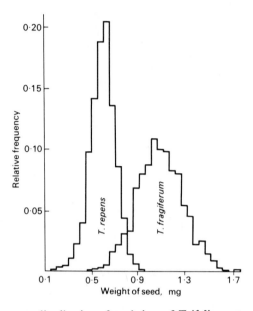

Fig. 6.16. Frequency distribution of seed sizes of *Trifolium repens* and *T. fragiferum*. (From Harper and Clatworthy 1963; courtesy of *J. exp. Bot.*)

4. GENETICAL CONTROL, SEED PRODUCTION AND TOTAL SYSTEM INTERACTION

Overall control of the growth pattern and behaviour is controlled by the genetic make-up of the individual, and in many instances at a level where phenotypic differences are not necessarily evident. Black (1960) showed that three strains of subterranean clover with leaf development at different heights above ground (Fig. 6.18) produced different light profiles resulting in marked interference between individuals in mixed populations (Fig. 6.19). Cavers and Harper (1966) using replicates of individuals of *Rumex crispus* and *R. obtusifolius* from six different locations showed an extremely variable response to two contrasting germination conditions (Fig. 6.20). Not only was there marked differences between locations but different individuals from the same site showed as much variation (Fig. 6.21). Although this polymorphism can be related to nannoclimate

(a)

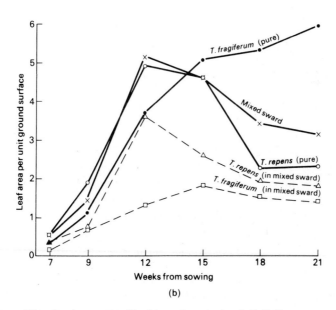

(b)

Fig. 6.17. The development of leaf area in swards of *Trifolium repens* and *T. fragiferum.* (a) 32 plants per 900 cm², (b) 64 plants per 900 cm². (**From Harper and Clatworthy 1963;** courtesy of *J. exp. Bot.*)

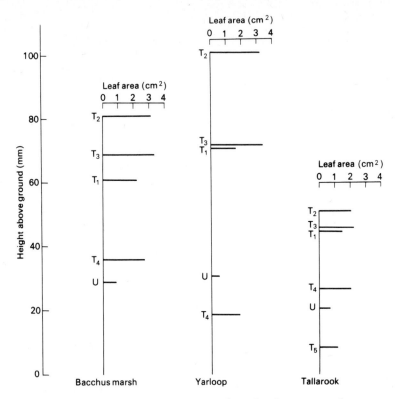

Fig. 6.18. Leaf development in three strains of subterranean clover grown under reduced light intensity. (From Black 1960; courtesy of *Aust. J. agric. Res.*)

of different parts of the inflorescence, or microclimate of different parts of the site, it is highly probable that there remains a residuum of variability directly related to genetical control. These two cases suggest for some species at least, that predictions as to population behaviour pattern will have to allow for the distribution of the gene population and their effect on seed germination and seedling development. There is now available a wide range of well-documented examples of ecotypic variation which has been very usefully collated by Heslop-Harrison (1964). In addition to relatively gross morphological differences (leaf lengths, hairiness, etc.) there is a wide range of physiological ecotypes. McKell *et al.* (1960) compared two races of *Dactylis glomerata* and showed very different responses to differing conditions of soil moisture stress, which correlated with their known distribution. Björkman and Holmgren (1963) in a study of photosynthesis efficiency in the sun and shade ecotypes of *Solidago virgaurea* found a more efficient use of weak light in shade ecotypes than

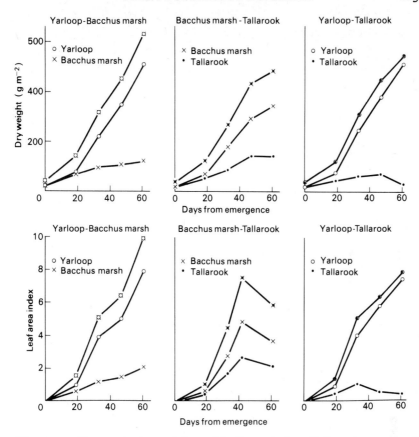

Fig. 6.19. Changes in dry weight and leaf area index of the components of mixed swards. (From Black 1960; courtesy of *Aust. J. agric. Res.*)

in those from open habitats. Conversely, clones from open habitats utilized intense light more efficiently. Similarly Mooney and Billings (1961) for *Oxyria digyna* showed that plants of northern populations had higher respiration rates at all temperatures than plants of southern alpine populations. Numerous additional examples are also available in the literature (see Heslop-Harrison 1964, for details). Although many of these ecotypic differences relate to geographically separated populations, similar though less extreme variability will occur in adjacent populations, where environmental pressure has been exerted on the gene population. Equally although the documented examples above do not relate directly to seed characteristics, they are all indirectly related and will control to a greater or lesser extent seed numbers, or quantity of stored metabolites and hence seed size.

Fig. 6.20. The percentage germination of seeds from ten different plants in each of six different habitats tested under contrasting germination conditions. Vertical lines: Light, alternating temperature; Black bars: darkness, constant temperature (20°C); White bars: (a–d) darkness, alternating temperature (10°–20°C), (e) and (f) darkness, constant temperature (20°C) interrupted by one brief exposure to daylight. (From Cavers and Harper 1966; courtesy of *J. Ecol.*)

The evidence from Cavers and Harper (1966) also indicates a marked seasonal effect on seed weight, harvested at three different times from individual plants of *Rumex cripus* and *R. obtusifolius*. This results in a corresponding variation of percentage germination (Fig. 6.22*a* and *b*).

There is a marked dynamic situation existing in the fruiting panicle. Variations of temperature and light induced by the detailed structure of the panicle and its particular orientation to the sun will not remain constant throughout the year. Both the continued development of the panicle itself, as well as the progressive change in angle of inclination of the sun's rays will profoundly affect nannoclimate at any one sample point. Where there is a temperature requirement for after-ripening of the seed, a further level of heterogeneity will thus be imposed on the seed population. In

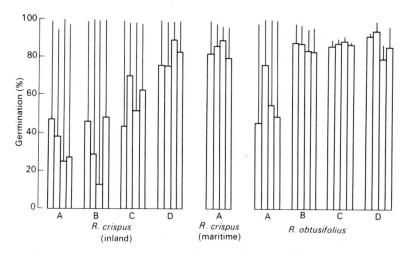

Fig. 6.21. The germination of seeds from different panicles of the same plant. Each block of histograms represents percentage germination of seed from four different panicles on a plant. Vertical lines: light, alternating temperature (10°–20°C); white bars: darkness, alternating temperature (15°–30°C). (From Cavers and Harper 1966; courtesy of *J. Ecol.*)

addition, seed weight which is known to subsequently affect the performance of the resultant seedlings is controlled by translocation of reserve metabolites up from the leaves. The availability of these metabolites is also a function of time of year.

The interactions outlined above relate only to documented examples and are not an exhaustive list. They result finally in the control of the reproductive efficiency of a species by a dynamic complex of factors. Harper (1960) has shown that the numbers of seeds produced is one resultant of intraspecific, or interspecific interference (see also Fig. 5.5, p. 97). Thus the feedback into the soil of seed *number* is also a variable, and the complexity of a system involving more than two species will be considerable. In addition, it would involve a number of populations of

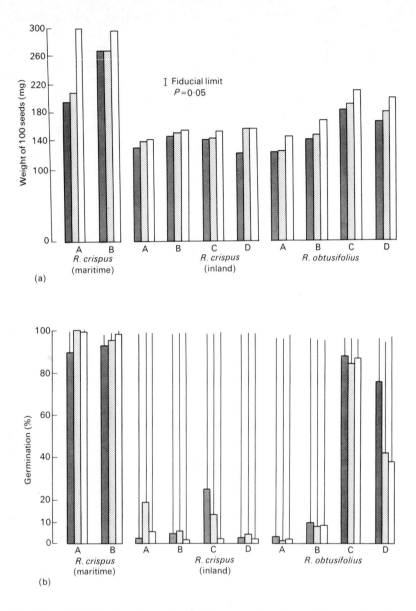

Fig. 6.22. (a) Variations in seed weight harvested at three different times from individual plants (A–D). Black bars, early harvest; stippled bars, medium harvest; white bars, late harvest. (b) Percentage germination of seeds harvested at different times from plants (A–D) as above. Histograms record germination under darkness and alternating temperature, vertical lines under light and alternating temperature. (From Cavers and Harper 1966; courtesy of *J. Eçol.*)

perennial species about which virtually no information is available. It is assumed that their establishment by seed is controlled by a similar system to that outlined above, but in an ecological situation the control of lateral bud production by intrinsic and/or extrinsic factors is unknown. There is an equal lack of information on biotic factors, other than generalized comments on the food habits of small mammals. It is clear, however, that even with the existing information we have on the control of population numbers in simple experimental situations, the variability of plant associations and the widespread occurrence of vegetational pattern, both become understandable.

7 The Poisson Series and the Detection of Non-Randomness

INTRODUCTION

IN addition to the detection of non-randomness in species distributions by correlation methods discussed in the last chapter, a number of approaches have been made to the non-random distribution of individuals of the same species. This, in part, can be accounted for in similar terms as correlation between species, but several other methods of detecting departure from randomness are available.

HISTORICAL

The earliest accounts of the non-random nature of the distribution of plants in a community are those of Gleason (1920) and Svedberg (1922) who independently showed that several species were markedly non-random. Svedberg's approach to the problem has since become one of the standard methods of detecting non-randomness in vegetation and consisted of relating the observed number of individuals per quadrat to the expected number derived from the Poisson series e^{-m}, $m\,e^{-m}$, $m^2/2!\,e^{-m}$, $m^3/3!\,e^{-m}$, $m^4/4!\,e^{-m}$, ..., where m is the mean density of individuals. Successive figures of the series give the probability of quadrats containing 0, 1, 2, 3, 4, ... individuals respectively, and the expected number of quadrats thus falling into each of these classes can be readily calculated. The nature of the non-randomness was termed as 'overdispersed' when individuals tended to be clumped together, and 'underdispersed' when individuals were scattered very evenly over the area. Overdispersion is thus characterized by large numbers of both empty quadrats and quadrats containing a large number of individuals, and similarly underdispersion by the majority of quadrats containing an intermediate number of individuals. Svedberg showed that dispersal of a number of species agreed satisfactorily with the Poisson series which indicated random dispersion, but also that several species showed either 'overdispersion' or 'underdispersion'. The terms 'overdispersed' and 'underdispersed' refer to the distribution curve of the data and *not* to the pattern of individuals on the ground. This has led to some confusion and it has been suggested by

Greig-Smith (1957) that the terms 'contagious' and 'regular' should re-place 'overdispersion' and 'underdispersion'. In a Poisson series the variance is equal to the mean and thus the ratio of these two values is equal to 1. Svedberg used this ratio as a measure of randomness; when the values were greater than 1 the distribution was assumed to be contagious, when less than 1 it was assumed to be regular.

TESTS OF SIGNIFICANCE

The use of variance:mean as an index of contagion in vegetation has since been used by a number of workers (Clapham 1936, Archibald 1948, Dice 1952, etc.) usually employing a significance test for the difference between the observed and expected variance:mean ratio (Blackman 1942) or a χ^2-test to compare the terms of the Poisson series with the observed data (Blackman 1935). In addition to the test of goodness of fit, and variance:mean ratio (sometimes termed the Coefficient of Dispersion (Blackman 1942) or relative variance (Clapham 1936)) several additional tests for non-randomness have been devised which are adequately de-scribed by Greig-Smith (1957) (see also Moore 1953, Ashby, 1935, David and Moore 1954, Whitford 1949) and are not discussed here. However, as Evans (1952) has pointed out the variance:mean ratio may give a widely different estimate of non-randomness from a χ^2-test of goodness of fit and it may be necessary to use more than one test for non-randomness. Evans gives the following hypothetical case with a mean and variance of 1·00 and hence a variance:mean ratio of 1 which indicates a random dis-tribution within the population:

Table 7.1 The observed and expected number of quadrats containing 0, 1, 2, ... individuals in a hypothetical case, where the variance and mean of the population equal 1

Number per quadrat	Observed number of quadrats	Expected number of quadrats
0	20	37·16
1	76	37·16
2	—	18·58
3	—	6·19
4	—	1·55
5	5	0·31
> 5	—	0·05

Obviously there is a considerable discrepancy between the observed and expected data ($\chi^2 = 63\cdot24$, $p \ll 0\cdot1$ per cent) and in general the χ^2-test of goodness of fit is a more reliable indication of non-randomness than the variance:mean ratio.

The method of testing departure from randomness is illustrated below. The data is taken from two communities, one consisting entirely of random individuals (Fig. 7.1) and one with hypothetical offspring grouped around the 'parents' (Fig. 7.2). The data has been obtained from randomly

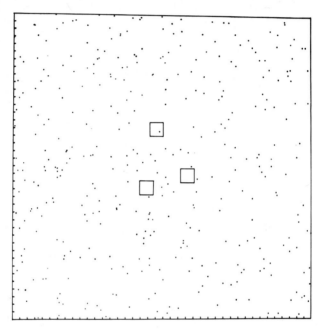

Fig. 7.1. A random 'community' with sample quadrats positioned by pairs of random co-ordinates.

placed quadrats located by pairs of co-ordinates, taken from tables of random numbers. Thus the first quadrat is located by co-ordinates 240.200 (the bottom-left hand corner of the quadrat), the second by co-ordinates 163.187 the third 179.241 and so on (a procedure identical with that used to position the 'individuals' (Fig. 7.1) in the 'community'.

1. RANDOM POPULATION (Fig. 7.1)

The data for 100 random samples is given below:

Table 7.2—The observed number of quadrats containing 0, 1, 2, . . . individuals taken from a random population

Number of individuals in each quadrat (a)	0	1	2	3
Frequency of occurrence in 100 quadrats (f)	46	34	14	6

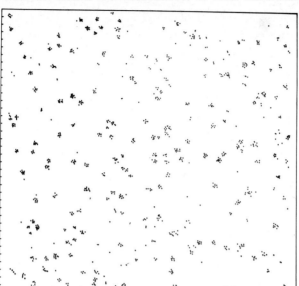

Fig. 7.2. A contagious distribution of individuals.

(a) χ^2 *goodness of fit*

$$\text{Mean density of the population, } m = \frac{Saf}{100} = 0 \cdot 8$$

Thus from the series $e^{-m}, m\,e^{-m}, m^2/2!\,e^{-m}, m^3/3!\,e^{-m}, \ldots$, the expected number of quadrats containing $0, 1, 2, \ldots$ individuals can be calculated:

$$e^{-m} = e^{-0 \cdot 8} = 0 \cdot 4493$$
$$\text{(see Greig-Smith 1957; Appendix B, Table 3)}$$

$$m\,e^{-m} = 0 \cdot 4493 \times 0 \cdot 8 = 0 \cdot 3594$$

$$\frac{m^2}{2!}\,e^{-m} = 0 \cdot 4493 \times \frac{0 \cdot 64}{2} = 0 \cdot 1438$$

$$\frac{m^3}{3!}\,e^{-m} = 0 \cdot 4493 \times \frac{0 \cdot 512}{6} = 0 \cdot 0383$$

Thus the expected distribution should be:

$$0 \cdot 4493 \times 100, \quad 0 \cdot 3594 \times 100, \quad \ldots \text{ and thus:}$$

Number of individuals per quadrat	0	1	2	3
Expected frequency	44·9	35·9	14·4	3·8
Observed frequency	46	34	14	6
Difference	1·1	1·9	0·4	2 2

The goodness of fit is calculated as the sum of the differences squared divided by the expected frequency:

$$= \frac{(1 \cdot 1)^2}{44 \cdot 9} + \frac{(1 \cdot 9)^2}{35 \cdot 9} + \frac{(0 \cdot 4)^2}{14 \cdot 4} + \frac{(2 \cdot 2)^2}{3 \cdot 8}$$

$$= 0 \cdot 0269 + 0 \cdot 1006 + 0 \cdot 0111 + 1 \cdot 2737$$

$$= 1 \cdot 4123$$

The χ^2-table is entered with 2 degrees of freedom (2 less than the number of terms used to calculate the χ^2 total) which shows a value of $1 \cdot 386$ when $p = 0 \cdot 5$. Thus the chance of this difference between the two sets of data arising fortuitously is 50:50 and we can regard the observed data as showing a very good fit with the expected series and the population sampled was randomly distributed. (When the chance of a difference between two sets of figures arising completely fortuitously, falls as low as $0 \cdot 05$ (1 in 20 or a 5 per cent level of significance) it is usually considered that such a level of odds necessitates some other hypothesis, and the difference can be safely regarded as 'real' rather than 'accidental').

(b) Variance : Mean ratio

From the data above:

$$N = 100, \qquad \bar{x} = \frac{S(x)}{N} = \frac{80}{100} = 0 \cdot 8$$

$$S(x)^2 = 144$$

$$(Sx)^2 = 6400$$

$$\therefore \frac{(Sx)^2}{N} = 64.$$

The variance of the population is given by

$$\frac{S(x)^2 - \frac{(Sx)^2}{N}}{N - 1} = \frac{144 - 64}{99} = \frac{80}{99} = 0 \cdot 8080$$

Thus the variance : mean ratio $= \dfrac{0 \cdot 8080}{0 \cdot 8000} = 1 \cdot 01$

The variance : mean ratio shows the population to have apparently some degree of contagion, and this difference from the expected ratio of 1 must be tested by a t-test:

Standard error of the variance : mean ratio is given by

$$\sqrt{[2/(N-1)]} = \sqrt{\tfrac{2}{99}} = 0 \cdot 1421$$

$$t = \frac{\text{Observed} - \text{Expected}}{\text{Standard Error}}$$

$$= \frac{0 \cdot 01}{0 \cdot 1421}$$

$$t = 0 \cdot 0704, \; p < 0 \cdot 9$$

Again this difference is not significant for such a difference could arise by chance very frequently, and the distribution of the population can be regarded as random.

II. CONTAGIOUS POPULATION (Fig. 7.2)

The data for 100 random samples is given below, the sampling procedure being identical to that used above:

Table 7.3 The observed number of quadrats containing 0, 1, 2, . . . individuals taken from a contagious population

Number of individuals in each quadrat	0	1	2	3	4	5	6	⩾7
Frequency of occurrence in 100 quadrats	47	6	5	8	5	6	7	16

(a) χ^2 goodness of fit

Mean density of the population, $m = 2 \cdot 44$

The Poisson series with this mean value is given by the probabilities:

$$e^{-m} = e^{-2 \cdot 44} = 0 \cdot 0872$$

$$m\,e^{-m} = 0 \cdot 2128$$

$$\frac{m^2\,e^{-m}}{2!} = 0 \cdot 2596$$

$$\frac{m^3}{3!}e^{-m} = \frac{2 \cdot 44^3}{6}\,0 \cdot 0872 = 0 \cdot 2111$$

$$\frac{m^4}{4!}e^{-m} = \frac{2 \cdot 44^4}{24}\,0 \cdot 0872 = 0 \cdot 1288$$

$$\frac{m^5}{5!}e^{-m} = \frac{2 \cdot 44^5}{120}\,0 \cdot 0872 = 0 \cdot 0628$$

$$\frac{m^6}{6!}e^{-m} = \frac{2 \cdot 44^6}{720}\,0 \cdot 0872 = 0 \cdot 0256$$

$$\frac{m^7}{7!}e^{-m} = \frac{2 \cdot 44^7}{5040}\,0 \cdot 0872 = 0 \cdot 0089$$

Number of individuals per quadrat	0	1	2	3	4	5	6	>7
Expected frequency	8·7	21·3	26·0	21·1	12·9	6·3	2·6	0·9
Observed frequency	47	6	5	8	5	6	7	16
Difference	38·3	15·3	21·0	13·1	7·9	0·3	4·4	15·1

$$\chi^2 = \frac{(38·3)^2}{8·7} + \frac{(15·3)^2}{21·3} + \cdots + \frac{(15·1)^2}{0·9}$$

$$= 470·334, \quad p < 0·001$$

The difference between the observed and expected numbers of occurrences is highly significant, the chances of this difference arising accidentally are very much greater than 1000:1.

(b) Variance:mean ratio

$$N = 100, \quad m = \bar{x} = 2·44$$
$$S(x)^2 = 1364·0$$
$$(Sx)^2 = 59\ 536·0$$
$$\frac{(Sx)^2}{N} = 595·36$$

$$\frac{S(x^2) - \dfrac{(Sx)^2}{N}}{N-1} = \frac{768·64}{99} = 7·7640$$

$$\therefore \text{ Variance:mean ratio} = \frac{7·7640}{2·44} = 3·182$$

Standard error $= 0·1421$

$$\therefore \quad t = \frac{2·182}{0·1421}$$

$$= 15·3554 \text{ with 99 degrees of freedom,} \quad p < 0·001$$

Again the probability of this difference arising by chance is very much less than 1000:1.

Various criticisms of the variance:mean ratio have been made, in addition to the case outlined by Evans (1952) described previously. Thus Jones (1955–6) suggests the interpretation of variance:mean ratios is very unreliable when the mean density of individuals is very high or very low, and Skellam (1952) has criticized this approach on the grounds that the success of the ratio as an indicator of non-randomness is dependent on the size of the quadrat used for sampling. This latter criticism is, in fact, applicable to all of the methods used in the detection of non-randomness of vegetation. This effect of size of quadrat can be very useful in gaining information on the scale of the non-randomness present and is discussed below. In general the χ^2-test for goodness of fit gives a reliable indication of the occurrence of non-randomness relative to the choice of quadrat size, provided very abundant and very rare species are not included in the investigation. Both these tests and others which have appeared are not applicable to distributions where an individual is not an obvious entity, and work on grassland for instance has accordingly been severely handicapped (see below).

Several attempts at deriving a method for detecting non-randomness in the distribution of tree species have been made. These have involved measures of point-to-plant and plant-to-plant distances.

Basically all the methods employing distance measures between individuals or between random points and individuals depend on the relationship between these measures in a random population. Thus when individuals are distributed at random, the ratio of mean distance from a random point to the nearest individual, to the mean distance between randomly selected individuals to its nearest neighbour, should be unity. This ratio will be greater than one when individuals are contagiously distributed and less than one when they are evenly distributed (corrected for the relative density of individuals). The significance of departure from expectation can be tested in a variety of ways.

CONTAGIOUS DISTRIBUTIONS

Following the demonstration that the observed distribution of individuals in a plant community did not fit a Poisson series, a considerable effort was made to find some mathematical series to which field data of this nature could be satisfactorily fitted. The type of function used in all cases involves parameters relating the distribution of individuals to random central points, around each of which a number of 'offspring' is scattered. Thus contagious distribution of individuals is related to the most obvious and likely causal factor—that of vegetative spread or heavy seeds from a parent individual the parents being distributed at random. Archibald (1948, 1950) found a satisfactory fit for many (though not all) species, to Neyman's contagious distribution and also to a similar distribution Thomas's Double Poisson. Similarly, Barnes and Stanbury (1951) found a satisfactory fit to Neyman's and Thomas's distributions using data taken from uniform deposits of china clay residues in the process of being colonized. On the other hand, Thomson (1952) found only one species out of three tested that fitted these distributions at all satisfactorily. Several other distributions have been suggested, based on the premise that contagion in vegetation is largely due to the morphology of the individual or the efficiency of seed dispersal mechanisms, and operates at one (small) scale only. Unfortunately the value of this approach is minimized by two considerations: (*a*) later work has shown that contagion in vegetation is due to a multitude of factors and is present on numerous scales in any one site; (*b*) the mathematical parameters employed to define the distribution and generate the series have no meaning ecologically, or at least it is impossible to relate known ecological factors to these parameters. Thus whilst the approach is of considerable academic interest to the mathematician it is of little use to the ecologist and it would appear that the complexity and variation in the distribution of individuals is of such an order that the formulation of a mathematical model is virtually impossible and it is necessary to fall back on rather simpler empirical approaches.

THE EFFECT OF QUADRAT SIZE ON THE DETECTION OF NON-RANDOMNESS

One of the criticisms made of the variance:mean ratio and goodness of fit as tests of non-randomness in vegetation, has been the dependence of its detection on quadrat size (Skellam 1952). On theoretical grounds it can be readily shown that in any contagious population the use of a Poisson series and tests of departure from it will show both random, contagious and regular distribution as the size of quadrat is steadily increased. This is illustrated below (Fig. 7.3) where sampling with quadrats *A* or *B* the

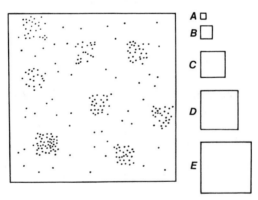

Fig. 7.3. The relationship between quadrat size and variance (see text).

population would probably show slight contagion, sampling with quadrat *C* very marked contagion (each quadrat would contain very many or very few). With quadrats *D* and *E* the distribution would appear to tend towards a regular distribution with all quadrats containing approximately the same number of individuals. Thus the most marked demonstration of contagion would be with the quadrat with an area approximately equal to the area of the clump.

Greig-Smith (1952*a*) gives data obtained from a series of artificial layouts of coloured discs, and some of the data is presented below (Table 7.4), which illustrates the apparent change of distribution of individuals in a population with increase in size of the sampling unit. For quadrat sizes 10, 15, 20 and 25 cm there is a marked indication of randomness; at 30 cm the relative variance (variance:mean ratio) increases sharply to a significant level which is maintained in the 35 and 40 cm quadrats, indicating the contagious nature of the artificial population.

(The relationship between quadrat size and variance can be readily appreciated from the hypothetical scheme above (Fig. 7.3). With the small quadrat *A* roughly equal proportions of the quadrats will contain high, intermediate and low densities of individuals. With quadrat *C* on the average high or low values will predominate and the variance accordingly will be high. Finally, a very large quadrat (*E*) will contain roughly equal numbers of individuals and the variance of the data will be very low.)

Table 7.4 Density data from Experiment 5 (Mosaic of irregularly shaped areas), showing the change of contagion detected with increase of quadrat size. (From Greig-Smith 1952*a*; courtesy of *Ann. Bot., Lond.*)

Quadrat size	Mean	Mean per 100 cm²	Variance: Mean ratio	t	p	χ^2	n	P
10 cm	0·07	0·07	0·9394	0·43	0·6–0·7	–	–	–
15 cm	0·18	0·08	1·0527	0·37	0·7–0·8	–	–	–
20 cm	0·29	0·07	0·9958	0·03	0·9	–	–	–
25 cm	0·39	0·06	1·1342	0·94	0·3–0·4	1·56	1	0·2–0·3
30 cm	0·68	0·08	1·3928	2·76	0·001–0·01	4·00	1	0·02–0·05
35 cm	0·79	0·06	1·5674	3·99	0·001	2·29	1	0·1–0·2
40 cm	0·97	0·06	1·6340	4·46	0·001	26·14	2	0·001

The χ^2 goodness of fit gives a closely comparable picture but it does not reach a significant level of departure from expectation until the 40 cm sampling quadrat. The approximate area of the 'mosaic' employed of high and low densities in this particular layout corresponds more or less with the 35–40 cm quadrat.

The importance of the relationship between quadrat size and detection of contagion lies in the fact that the actual scale at which clumping occurs can be detected merely by resampling the population several times with different sizes of quadrats. Since non-randomness may thus be demonstrated in vegetation where a close visual inspection does not reveal any obvious sign of clumping, this is an important step forward The mere demonstration of non-randomness in vegetation is of very limited interest but once information is available as to the scale or scales at which this non-randomness occurs, it becomes possible to relate some environmental factor or factors to scales of the detected non-randomness.

THE ANALYSIS OF A CONTIGUOUS GRID OF QUADRATS AND THE DETECTION OF PATTERN

The development of the analysis of a grid of quadrats is a logical step following the demonstration of the dependence of the detection of non-randomness on quadrat size. Instead of throwing a range of quadrat sizes over an area, a grid of contiguous quadrats is laid out and enumerated, the increasing 'quadrat' sizes then being built up by blocking adjacent quadrats in pairs, fours, eights, etc. An analysis of variance of the data is then carried out, the variance being partitioned between the different block sizes. In the graph relating the mean square (variance) to block size the different scales of pattern appear as peaks at a block size corresponding to the mean area of 'clump'. As was pointed out above, this approach is normally used where no visual contagion (pattern) is detected and the term 'clump' is, more strictly speaking, an area where a species is at a slightly higher or lower density than in the surrounding area. This slight difference in density is not apparent on a visual inspection.

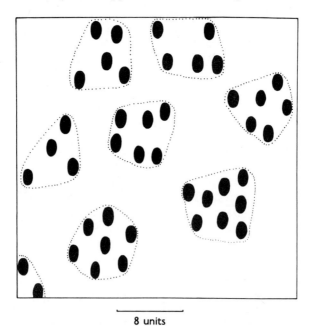

8 units

Fig. 7.4. An artificial 'community' with two scales of pattern (see text).

An example of such an analysis is given below with data taken from an artificial layout of clumps 2 units square grouped into a second scale of pattern at 64 units (Fig. 7.4). The analysis of variance table (Table 7.5) and graph of mean square against block size (Fig. 7.5) shows very clearly the two scales of pattern present in the layout as a double peak.

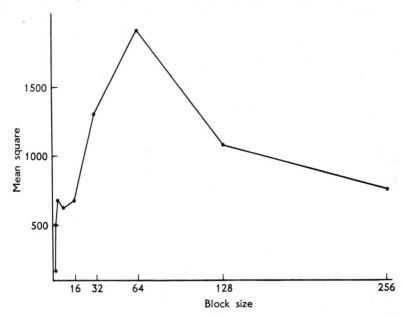

Fig. 7.5. Mean square (variance)/block size graph with a double peak representing two scales of pattern (see text).

Table 7.5 Analysis of variance for data taken from an artificial layout with two scales of pattern

Block size (Ns)	(Sx^2)	$\dfrac{(Sx^2)}{Ns}$	Sum of squares	Degrees of freedom	Mean square (variance)
1	483 867·45	483 867·450	84 709·815	512	165·449
2	798 315·27	399 157·635	129 738·643	256	506·792
4	1 077 675·97	269 418·992	88 065·841	128	688·014
8	1 450 825	181 353·151	40 492·551	64	632·696
16	2 253 769·61	140 860·600	21 728·739	32	679·023
32	3 812 219·57	119 131·861	21 025·536	16	1314·096
64	6 278 804·81	98 106·324	15 293·160	8	1911·645
128	10 605 382·63	82 813·165	4 337·279	4	1084·320
256	20 089 826·77	78 475·886	1 537·425	2	768·712
512	39 392 492·49	76 938·461	93·431	1	93·431
1024	78 689 318·49	76 845·030	—	—	—

The original method entailed using density as a measure of abundance but the approach was modified and extended (Kershaw 1957) to enable percentage frequency or cover data to be used when the technique was being applied to grassland communities, etc. Frequency can be very readily utilized in this type of analysis, the basic unit of the grid being a sub-divided quadrat instead of a simple quadrat. Presence or absence is recorded for each sub-division of the basic grid unit (25 sub-divisions are usually sufficient) and a percentage frequency figure calculated for each quadrat in turn of the whole grid. The frequency data are then blocked up as before and analysed, the graph mean square/block size peaking at the mean area of 'clump'. The conversion of the technique to cover data necessitates a change of approach in that instead of detecting the mean *area* of a 'clump' a line of cover readings is taken and analysed giving the mean *dimension* of a scale or scales of pattern. A Perspex frame has been designed to enable contiguous cover readings to be taken at intervals as small as 1 cm along a transect (Fig. 7.6). Thin needles are lowered, or optical measure taken (by sighting through the two sets of holes on to the ground layer) and recorded, down one edge of the frame. The frame is then pivoted and the readings continued along the next row of holes that have thus been brought into line. Each set of five adjacent readings are grouped together and expressed as a percentage cover reading. The cover readings are then blocked up and analysed, the peaks in the graph representing the mean *dimension* of a scale of pattern.

Originally it was thought necessary to transform the data before analysing it but it appears that this is not essential (Greig-Smith 1961a) unless tests of significance are to be made. The significance of any particular peak represents a difficulty that was finally overcome by a subjective approach. Under non-random conditions the use of a variance ratio test is invalidated and it was therefore necessary to fall back on the consistent appearance of a peak in a set of replicates. Thus if in a series of replicate samples two scales of pattern are detected at the same or closely similar block sizes in all the analyses, they can be taken as certain departures from randomness at these two scales. Thompson (1958) has considered the statistical implications of the method and also suggests repeated sampling of a population will distinguish chance effects from steady trends inherent in the community. The general reliability of this approach has been tested by trials on artificial layouts of different types (Kershaw 1957). In general as long as the basic unit is smaller than the smallest scale of pattern to be detected, the abundance of the species being investigated is not too low and some degree of replication is obtained, the results are reliably consistent (see also Greig-Smith 1961a).

It is more usual to adopt the dimensional approach even when using density data, since it is easier to obtain rapidly a clear picture of the dimension of the scales of pattern present with four transects of quadrats than it is to add more rows of transects to build up a complete grid. Numerous examples are available of various patterns in vegetation that have developed in response to certain factors and it is clear that the technique is more than sufficiently sensitive to detect very faint pattern.

The use of this approach to non-randomness in vegetation does enable a detailed approach to be made to the causal factors underlying the pattern (see below). When a pattern is very slight it is likely that one or relatively few environmental factors will be controlling it. Conversely, if a

Fig. 7.6. A point quadrat for taking contiguous cover readings along a transect. (From Kershaw 1958; courtesy of *J. Ecol.*)

pattern is very marked, it is probable that a number of environmental factors are involved. In the former case it is often possible to make an approach to the causal factor involved, in the second instance the factors are often so numerous as to defeat any attempt at elucidating their effects and interactions. Similarly where a pattern is clearly visible without recourse to detailed statistical analysis, the complexity of the environmental background of such patterns is too great to enable a clear insight into the detailed mechanism. Thus a study of pattern is of considerable use in studies of interactions between the environment and vegetation and also between individual and individual. Pattern can similarly be expressed as a level of association between species (see an earlier section) and in fact the trend of association between species is often of considerable use in gaining an understanding of the causal factors of the pattern. Thus once two species are positively or negatively related in a community, the population will be non-random and the pattern thus present can be expressed either as an analysis of variance or as a correlation coefficient.

The relationship between quadrat size and the trend of detected association discussed in an earlier section, is very marked in the variance analysis.

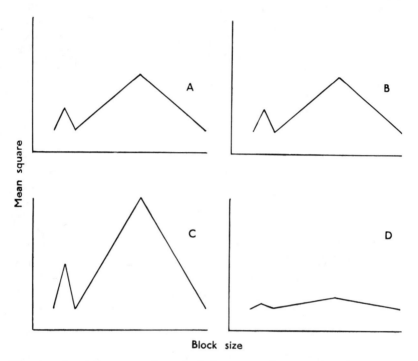

Fig. 7.7. *A* and *B* represent hypothetical patterns of two species when analysed separately. If analysed jointly their pattern will be as *C* if they are positively associated and as *D* if negatively associated.

Correlation coefficients similarly can be calculated for different block sizes and the change of trend of association related to the scales of pattern of the species. A single approach to this aspect of non-randomness can be made by considering the covariance between two species and analysing it in a similar way to variance (Kershaw 1960). Thus if two species *A* and *B* are not related in any way then Var A + Var B = Var $(A + B)$. Conversely if they are associated in some way, the observed variance will be different from the expected variance. This is conveniently illustrated by considering two species *A* and *B* having two scales of pattern as shown above in Fig. 7.7 *a* and *b*. If they are associated together positively then the analysis of the joint data of *A* and *B* (i.e. the original data of *A* added to the original data of *B*) will show an increase of variance over and above that expected on theoretical considerations (Fig. 7.7c). Similarly if *A* and *B* are negatively related, the joint analysis of *A* and *B* will show a decrease of variance compared with that expected (Fig. 7.7d). The difference between the

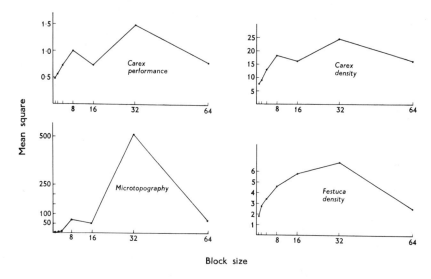

Fig. 7.8. The scales of pattern in *Carex bigelowii*, *Festuca rubra* and micro-topography in an Icelandic *Rhacomitrium* heath (from Kershaw 1962; courtesy of *J. Ecol.*)

observed mean square/block size graph for *AB* and that expected (i.e. the mean square/block size graph of *A* plus that of *B*) is a measure of the co-variance between *A* and *B* at different block sizes. An example is given above showing the relationship between *Carex bigelowii* and *Festuca rubra* in *Rhacomitrium* heath in Iceland (Kershaw 1962) (Fig. 7.8). The micro-topography of the area shows two scales of pattern at block size 8 and 32 (equivalent to 80 cm and 32 m). The density and performance data (ex-pressed as mean leaf length per quadrat of *Carex*) show an identical pattern which immediately suggests a correlation with the microtopo-graphy of the area. The variance analysis of the density data for *Festuca rubra* is not as clear cut, apparently lacking the primary peak at block size 8 but having a marked peak again at block size 32. The co-variance analyses (Fig. 7.9) confirm the suspected relationship between micro-topography and the distribution of *Carex*. Both *Carex* density and *Carex* performance are obviously positively correlated with microtopography and it is interesting to note the amplitude of the peak for performance/ microtopography is greater than density/microtopography. This is a clear

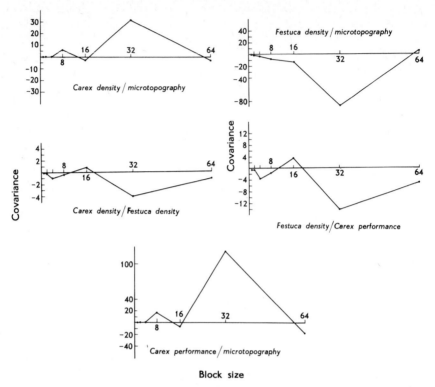

Block size

Fig. 7.9. The relationships between *Carex* density, *Carex* performance, *Festuca* density and microtopography, expressed as covariance (see text). (From Kershaw 1962; courtesy of *J. Ecol.*)

indication that the performance of *Carex* is affected more markedly by microtopography than is the density. *Festuca* and *Carex* are shown to have a markedly negative correlation (Fig. 7.9) and again the performance of *Carex* is more noticeably affected by *Festuca* than the density, which suggests the two species are competing actively. Finally the relationship between *Festuca* and microtopography has a markedly negative correlation, as would be expected. The data were obtained from eight replicated lines of 10 cm quadrats, and the analyses accordingly show the mean dimension of scales of pattern. It is interesting to see how the co-variance analysis has picked out the level of interaction between the species concerned and the microtopography of the area, which enables the relative vigour of the species and also the relative effect of the environment (in this instance, microtopography) on the two species to be assessed.

8 The Causal Factors of Pattern

PATTERN in vegetation is the spatial arrangement of individuals of a species; correlation methods are usually employed to examine the detailed relationships existing between different species, but clearly both are different aspects of a general phenomenon. It is impossible to discuss one without considerations involving the other and, accordingly, the causal factors relating to pattern overlap, to a considerable extent, those factors relating to correlation discussed in Chapter 5. This separation of 'pattern' and 'correlation' is artificial, but has been made for the sake of clarity.

Since the initial demonstration by Gleason (1920) and Svedberg (1922) of patchy distribution of individuals in an apparently uniform area, a considerable effort has been made to detect non-randomness in vegetation. Most authors concluded that vegetative spread and heavy seeds were the two most likely factors which cause such an aggregation of individuals. Ashby (1948) endorsed these views but pointed out that 'overdispersion is present in species where there is no obvious biological reason for it'. He suggested that an overdispersed species may produce a secondary overdispersion among other species of the community, and this could possibly account for the otherwise inexplicable pattern (discounting obvious soil heterogeneity). Goodall (1952b) and Greig-Smith (1952a) both suggested that cyclical regeneration in patches as postulated by Watt (1947a) could produce pattern in vegetation, but there still remained numerous cases where an explanation was difficult. Following the development of techniques of pattern analysis (Greig-Smith 1952a, 1961a; Kershaw 1957, 1960), it is apparent that a very wide range of pattern scales are present in vegetation. Some of these patterns probably reflect a correspondingly wide range of variation of some factors of the environment. Sometimes it has proved impossible to understand all the factors responsible for a certain pattern and often a scale of pattern is related to one of the more obvious or more easily measured of a group of closely inter-related environmental factors.

1. *MORPHOLOGICAL PATTERN*

Where a species has a densely tussocked form, or where the basal leaves form a rosette, then the smallest pattern present in an analysis will be a measure of the mean size of an individual. In these cases it is clear that the morphology of a species will produce small scale pattern and, furthermore, this scale of pattern will be imposed on other species present. An example of apparent spatial exclusion is given by Kershaw (1958) for *Agrostis tenuis*, which showed a marked scale of pattern at 3·2 m. *Dactylis glomerata* and *Lolium perenne* also showed this same scale of pattern but were negatively associated with *Agrostis*, and thus appeared to be spatially excluded. In fact the distribution of these species was controlled in the area mainly by the depth of soil, *Agrostis* growing more densely on the shallower areas of soil, and the other species on the areas of deeper soil. Thus pattern which would appear at first sight to be due to spatial exclusion of one species by another, may in fact be controlled by independent factors of the environment.

Where the species under consideration has an extensive rhizome system, several scales of pattern are usually evident as the result of the characteristic system of branching and leaf-production of that species. Phillips (1954a) showed for *Eriophorum augustifolium* three scales of pattern that could be attributed to morphology (Fig. 8.1). The morphology of *Eriophorum* gives a characteristic grouping of aerial shoots (Fig. 8.2)

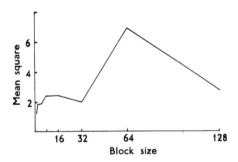

Fig. 8.1. Three scales of pattern in *Eriophorum augustifolium* represented by three peaks in the graph mean square/block size. (From Phillips 1954a; courtesy of *J. Ecol.*)

separated by a length of rhizome which is equivalent to a year's increment of growth. The double primary peak is due to the grouping of shoots round the parent shoot from backwardly directed short rhizomes giving a peak at block size 2 (200 sq. cm) and a larger scale of grouping due to longer rhizomes directed forwards and giving a peak at block size 8 (mean area 800 sq. cm). The large peak at block size 64 reflects the mean size of a complete individual (mean area 0·64 sq. m). A similar pattern series was detected in *Trifolium repens* (Kershaw 1959) with three scales of pattern

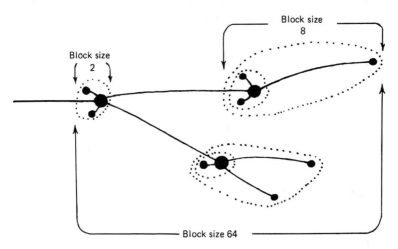

Fig. 8.2. Diagrammatic representation of the three scales of morphological pattern in *Eriophorum augustifolium*.

(Fig. 8.3) equivalent to first and second order branches and to the complete stolon system of the plant (Fig. 8.4). Numerous other examples of morphological pattern have been described (Agnew 1961; Anderson 1961*a*, 1961*b*; Greig-Smith 1961*a*, 1961*b*; Kershaw 1958, 1959; Kershaw and Tallis 1958), and in general, scales of pattern at small block sizes may be attributed to the morphology of the species.

The pattern due to the morphology of the rhizome can give useful information as to the level of performance of a species in different areas. It is clear that a measure of performance such as the mean annual extension of a rhizome system gives a good indication of the vigour of a plant and one which would be markedly controlled by the environment. Thus a series of pattern analyses from different plots over the range of distribution of a species should show a progressive shift of peaks which would reflect the changing morphology. Phillips (1954*a*) demonstrated such a variation of morphology in *Eriophorum* growing in different plant communities; the scales of pattern reflecting the morphology of the rhizome were very much smaller where the habitat was less favourable. (Fig. 8.5 and compare Fig. 8.1 above.) Anderson (1961*a*) and Kershaw (1959) discuss similar effects in *Vaccinium, Calluna, Pteridium* and *Trifolium* and it is apparent that valuable information can be obtained from comparisons of this sort.

The analysis of morphological pattern can also be related to the age of a rhizome system, as the pattern and size of branches vary with age and correspondingly alter the trend in a series of pattern analyses. The mor-

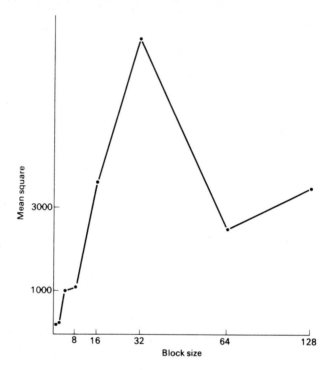

Fig. 8.3. Three scales of pattern in *Trifolium repens* shown by three peaks, at block size 1, 4 and 32, on the graph mean square/block size.

phological pattern of *Agrostis tenuis* colonizing a derelict sand pit shows such a progressive change (Kershaw 1958). The patches of sampled *Agrostis* varied in diameter from young clumps a few centimetres in diameter to large circular patches. The scales of pattern detected corresponding to the diameter of the patch are given below (Table 8.1). The three scales of pattern (block sizes 2, 8 and 32, equivalent to 2, 8 and 32 sq. in.) in the smallest clump correspond to first order branch and second order branch and to the complete rhizome system respectively. As the clumps increase in diameter the size of the complete rhizome system progressively increases, but the primary and secondary scales are simultaneously maintained. At the largest clump size investigated the maximum scale of pattern and the smaller scales of pattern are maintained and there appears a fourth scale of pattern which is caused by the primary branch systems separating away from the parent rhizome by decay of the older parts and appearing as established individuals. These 'fragments' will presumably then follow the same cycle of events.

Thus the morphological pattern, though usually expected at the smaller block sizes, is of considerable interest and can be often helpfully used to demonstrate the relationship between the morphology of the individuals and the environment, or conversely a progressive change in morphology with time. It should be pointed out that not all small-scale patterns are necessarily morphological patterns, and in fact pattern at block sizes 1–2

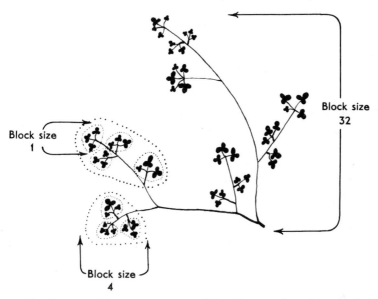

Fig. 8.4. Diagrammatic representation of the three scales of morphological pattern in *Trifolium repens*.

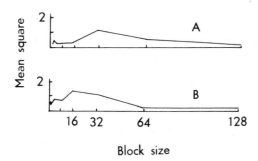

Fig. 8.5. The variation of pattern scales of *Eriophorum augustifolium* in two communities, *Calluna vulgaris*, *Erica cinerea* (*A*) and *Trichophorum caespitosum* (*B*). (From Phillips 1954*a*; courtesy of *J. Ecol.*)

Table 8.1 The scales of pattern present in analyses of tiller density from clumps of *Agrostis* of varying diameter. (From Kershaw 1958; courtesy of *J. Ecol.*)

Clump width (inches)	Density of tillers	Peaks (square inches)			
38	1·26	2	8	32	
42	1·42	2	8	64	
45	1·56	2	8	64	
51	1·74	2	8		128
54	1·72	2	8–16		128
90	1·90	1	8	32	128

(mean dimensions 5–10 cm) has in several instances been attributed to other intrinsic properties of the plant, or of the community or environment (see below).

Conversely morphological patterns are not always restricted to the smaller block sizes; they can equally be detected at the larger block sizes.

Chadwick (1960) describes an interesting historically determined pattern in *Nardus stricta*, which is best described as a large-scale morphological pattern. Two scales of pattern are detected, one at block size 2–4 (10–20 cm) caused by the morphology of *Nardus*, and a secondary peak at block size 32 (160 cm). The consistency of the secondary peak (Fig. 8.6) is surprising, and it would seem it is therefore not correlated with an environmental factor which would in fact fluctuate from site to site. Chadwick interprets this scale of pattern as the size of 'clone', reflecting the changes of rhizome growth since the grazing management of the Welsh mountains was changed. Thus with the advent of rationing in the first world war and the demand for smaller joints of meat the hill farmers changed over from old wethers which had effectively controlled the spread of *Nardus* by grazing, to breeding ewes which are much more selective in their grazing and do not touch *Nardus* unless more palatable species are absent. In response to this change the *Nardus* was able to spread and has actively done so ever since. Examination of the rhizome morphology shows a multidirectional growth, the old rhizome dying away behind and giving rise to a number of separate clumps each of which continues (as an individual) the centrifugal growth of the colony. The annual increment of growth is about 2 cm which over a period of 40 years would result in a scale of pattern of diameter of 160 cm equivalent in fact to the pattern at block size 32.

Austin (1968) examining the pattern in *Zerna erecta* growing on the North Downs in Kent, found very marked levels of contagion at the larger block sizes ranging from block size 32 to 128. The results were interpreted in an analogous way to Chadwick's *Nardus* data. Using an extensive series of turf dissections together with a vector analysis of the data in each block

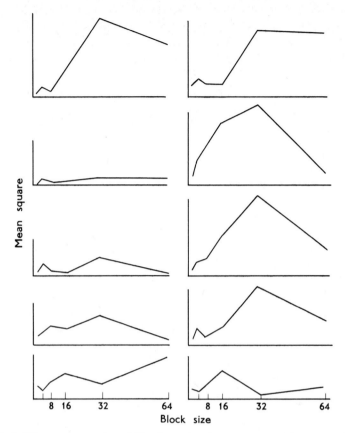

Fig. 8.6. The pattern scales of *Nardus stricta* in a number of upland grassland sites (see text). (From Chadwick 1960; courtesy of Blackwell Sci. Pub.)

dissected, Austin demonstrated a pattern of rhizome orientation consistent with growth from centres of colonization (Fig. 8.7). The spreading rings could be variable in size and since the length of 'intercept' of the randomized sample transects could also vary markedly, the observed variation of this pattern scale is understandable.

2. *ENVIRONMENTAL PATTERN*

The effects of major discontinuities of the environment on vegetation are usually well marked and, though such reactions of the vegetation result in patterns in the true sense of the word, these major effects, often on a large scale, are obvious as a change of floristic composition. Such a pattern is readily visible without recourse to quantitative method and calls for little comment. It has become increasingly evident that extremely small variations of an environmental factor (or factors) operating over quite large areas will produce a corresponding variation in the vegetation. Such patterns usually become apparent at the medium and larger block sizes in any particular area. Thus environmental pattern is developed in vegetation in response to a general and overall variation of one of the major environmental factors and produces a pattern of density distribution. The first demonstration of environmentally determined pattern showed the close relationship between the pattern of *Agrostis tenuis* and soil depth in an upland grassland area which was mentioned above (Kershaw 1958). The associated species *Dactylis*, *Lolium* and *Trifolium* were more abundant on the deeper areas of soil while conversely *Agrostis* was more dense on the shallower soil and their pattern, though on the same scale, was in fact an inverse pattern. The mean rooting depths of these species under normal conditions suggests a possible explanation of this pattern: *Agrostis* normally produces a shallow root system whilst the other species root at much greater depths. Thus the limitation of root development in *Dactylis*, etc., seriously affects the capabilities of these species and accordingly they are only likely to survive on the areas of deeper soil (Kershaw 1959). The effect of a small variation in soil depth on the distribution of species in the area was initially somewhat unexpected, but it now seems highly likely that numerous other soil factors will be shown to have a measurable effect on the distribution of species within an area which at first sight appears to be uniform vegetation.

Brereton (1971) examining the population structure of *Salicornia* and *Puccinellia* in salt marshes in N. Wales concluded that it was basically related to soil texture. Variation of soil texture in a spatial pattern produces a correlated pattern of soil moisture level which, in turn, controls species survival. Soil moisture level also affects soil plasticity, aeration and salinity, all of which affect the establishment of seedlings and their subsequent performance.

Anderson (1961a and 1961b), investigating the pattern scales in *Pteridium* and *Vaccinium myrtillus*, examined a number of edaphic factors including soil depth, pH, total exchangeable bases and oxygen diffusion rate. Of these factors, oxygen diffusion rate showed a strong correlation with the pattern of distribution of both *Pteridium* (Fig. 8.8) and *Vaccinium*. Hall (1967, 1971) examined the pattern scales and their interactions in *Brachypodium pinnatum*, *Festuca rubra*, and *Bromus erecta* on the S. Downs in Kent. He revealed an extremely complex situation with several marked pattern correlations with soil nitrogen, iron and boron. The pattern data for *Festuca* is shown below (Fig. 8.9) together with the marked interactions with soil iron and boron. Both these plant trace-nutrients are known to be usually of low availability in calcareous soils

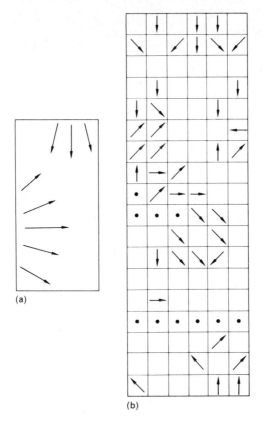

(a)

(b)

Fig. 8.7. (a) Proposed orientation of stolons in the turf (b) observed average orientation of stolons after vector analysis within each grid unit. (From Austin 1968; courtesy of *J. Ecol.*)

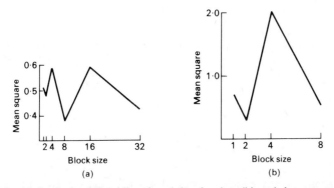

(a)

(b)

Fig. 8.8. (a) Analysis of *Pteridium* frond density data (b) and the corresponding soil oxygen diffusion data. (From Anderson 1961; courtesy of *J. Ecol.*)

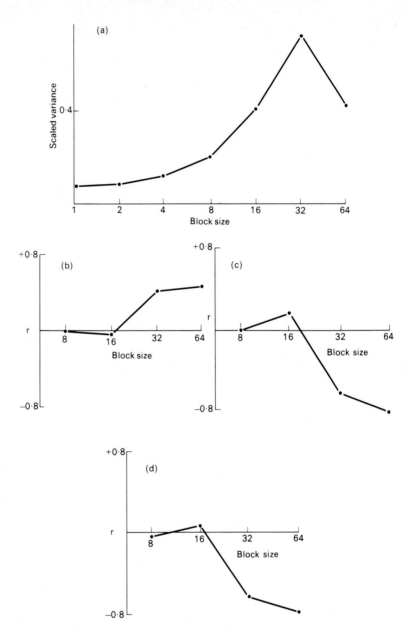

Fig. 8.9. (a) The scale of pattern in *Festuca* with (b) the correlated pattern of soil iron and (c) the correlated pattern of soil boron. (d) The correlation of pattern scales between soil iron and soil boron levels. (From Hall 1967; Ph.D. Thesis, London.)

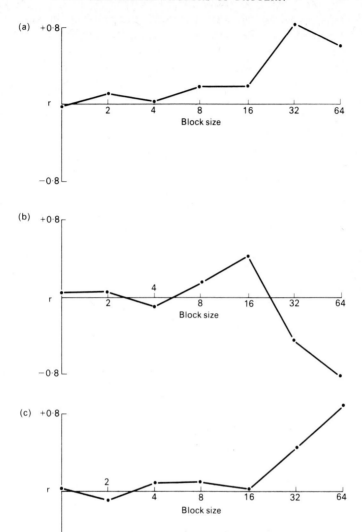

Fig. 8.10. The correlation of soil depth and pattern scales in (a) *Festuca*, (b) *Bromus*, (c) *Brachypodium*. (From Hall 1967; Ph.D. Thesis, London.)

(Russel 1961), and it is understandable that they are involved in the control of the pattern of at least the common chalk grass species. Hall also shows an overall control of pattern by soil depth with a resultant species interaction which is very clear cut (Fig. 8.10). Clearly soil factors are of fundamental importance in the control of pattern scales, but it is only recently that specific soil parameters have been correlated with particular pattern scales. This largely results from the difficulties of analysing soil properties in large numbers of replicates. In many situations soil properties correlating with pattern scales will represent a complex of individual parameters. Where pattern is related to obvious surface micro-topographical variation a similar complex of interacting factors will also operate. For example Greig-Smith (1961b) showed that the pattern of soil level formed of alternating hollows and mounds was correlated with the distribution of several dune slack species. Similarly, Kershaw (1962c) obtained a relationship between the micro-topography of a *Rhacomitrium* heath in Central Iceland and the distribution of *Carex bigelowii* and *Festuca rubra* (see above). In some reported investigations there is a marked pattern at the largest block size caused by an almost imperceptible slope in the area sampled and such a large and intense scale of pattern may often obscure smaller patterns due to other causes (see Greig-Smith 1961a). Similar, but more obvious, patterns have been reported by Billings and Mooney (1959), Billings and Mark (1961) and Warren Wilson (1952).

Thus it seems likely that drainage, availability of water, leaching, nutrient supply and pH would tend to vary together with changes in micro-topography and it would be difficult, if not impossible, to decide which factor or group of factors was directly controlling the distribution of a species. However, it is suggested that a slight fluctuation of pH or ion concentration would result in a detectable vegetational pattern, and it is interesting that Pigott has demonstrated considerable pH gradients over very short distances in the Burren Peninsula in Ireland. Snaydon (1962) has shown a similar steep pH gradient and also a gradient of calcium phosphate and potassium concentration, varying by a factor of up to 3 in a distance of only 2 ft (60 cm). It seems possible that similar or even steeper gradients will prove to be a general characteristic of undisturbed soils, producing the corresponding variations in the vegetation (see also Chapter 5). Thus the demonstration of a correlation between soil oxygen diffusion rate, and concentration of soil iron or boron (see above) are only three of the probable soil factors controlling pattern. Undoubtedly micro-gradients of other parameters will in the future be demonstrated as important in the control of pattern.

Other environmental parameters are of equal importance in controlling pattern scales. Kershaw and Rouse (1971b) examined the pattern in the homogeneous and often pure stands of the ground lichen *Cladonia alpestris*, in spruce-lichen woodland in N. Ontario. It had been shown previously that the lichen mat acts as an extremely efficient mulch, the moisture content of the soil layers underlying a sample quadrat effectively

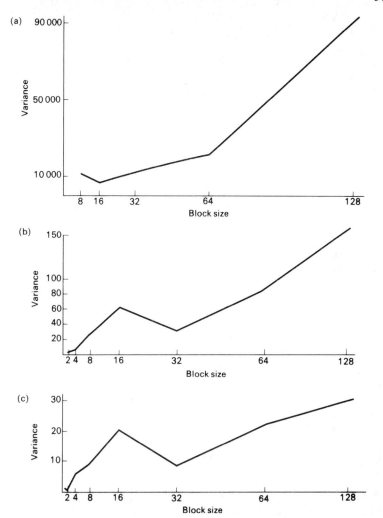

Fig. 8.11. The pattern analysis of the ground layer of uniform spruce-lichen woodland. (a) Depth of lichen mat; (b) *Ledum groenlandicum*; (c) soil moisture. (From Kershaw and Rouse 1971; reproduced by permission of the National Research Council of Canada from *Can. J. Bot.*, **49**, pp. 1389–1399).

denoting the relative amounts of integrated net radiation at that point, over the preceding period of weeks. Furthermore, since the metabolism of a lichen is entirely dependent on it being wet or at least moist, the relative drying rate of the lichen canopy will also control the growth pattern of the lichen. Pattern analysis of the depth of lichen mat, soil moisture (as a measure of incoming energy and hence drying rate of the lichen) and per cent cover of an associated woody shrub species *Ledum*

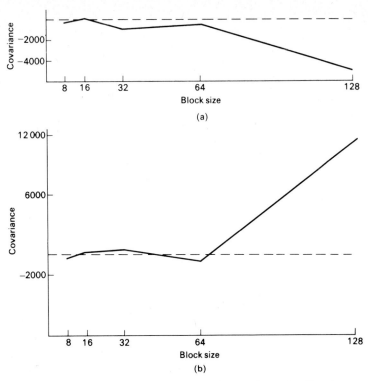

Fig. 8.12. Covariance analysis of the pattern interactions of soil moisture with (a) *Ledum groenlandicum*; (b) depth of lichen mat, in spruce-lichen woodland. (From Kershaw and Rouse 1971; reproduced by permission of the National Research Council of Canada from *Can. J. Bot.*, **49**, pp. 1389–1399.)

groenlandicum, are given above (Fig. 8.11). The covariance data (Fig. 8.12) show a marked positive relationship between depth of mat and soil moisture and a negative relationship between *Ledum* and soil moisture. The homogeneous lichen surface in fact is a mosaic of differential growth rates induced by a mosaic of differential rates of drying of the lichen surface. These rates in turn are related to a complex of tree shadows changing throughout the day and year as the sun alters its position. The existence of this gradient and its effectiveness was demonstrated by measurements of the drying rate of the lichen surface, at varying distances into an open area of woodland (Fig. 8.13). Yarranton and Green (1966) have shown two marked pattern scales in four lichen species growing on rock surfaces. The primary pattern scale is related to the angle of slope of the rock surface and the secondary pattern scale to vertical position on the cliff. Certainly the secondary pattern scale represents an additional example of micro-climate control of pattern scales, and it is probable that

(a)

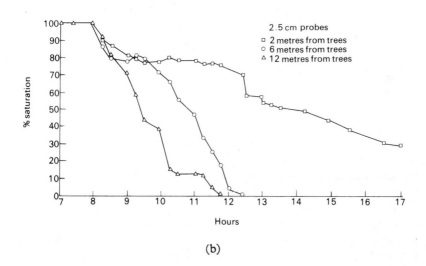

(b)

Fig. 8.13. Drying rate of the lichen canopy at (a) the surface, (b) 2·5 cm down on three sites, 16 m (1), 7 m (2), and 2 m (3) into an open area in spruce-lichen woodland. (From Kershaw and Rouse 1971; courtesy of *Can. J. Bot.*)

in the future other microclimate parameters will be shown to control patterns of species distribution (see also Barclay-Estrup 1971, p. 78).

3. SOCIOLOGICAL PATTERN

The term 'sociological pattern' covers a range of pattern units which are the products of several inter-related causal factors operating at small block sizes (up to a pattern scale of 80 cm dimensions in grassland, for example). These factors are partly due to intrinsic properties of the plants themselves and partly due to the micro-environment. Thus sociological pattern is a product of the interaction of species on species and individual on individual which may or may not be directly modified by micro-environment.

These small-scale sociological effects often take the form of a mosaic of patches of different density levels dynamically related to each other in time and space. The causal factors of sociological pattern are not only dependent on the competitive ability of an individual (which may also be, in part, dependent on the micro-environment), but also on the possible presence of toxins exuded by an individual, and its age. Several of these relationships can be examined by correlation methods and have been discussed previously (Chapter 5). However, they are usually evident in studies of pattern and are included here for completeness.

Almost every effect of one species on another (excluding parasitism) is in the form of a modification of the environment and it is possible that the environmental pattern which limits the distribution (in terms of density) of a species was in the first place imposed on the environment by another species. Zinke (1962) investigated the modification of the environment under *Sequoia gigantea*, *Pinus contorta* and other species. These modifications were expressed by pH, nitrogen content and exchangeable bases in the surface mineral soil. There was a marked radial relationship between these soil factors and tree cover (Fig. 8.14) and it is readily understood how such a pattern of environmental factors would impose a corresponding pattern of ground vegetation detectable for several years after the removal of the tree cover. The steady accumulation of leaf litter and corresponding increase in soil nitrogen is a well marked and expected result of tree establishment in a given site. It is very likely that similar less obvious gradients are developed in relation to all perennial species and a mosaic of micro-environmental variations will be rapidly developed which will subsequently influence the establishment of other species or individuals in the area. Accordingly some small-scale patterns, which strictly speaking should be grouped under 'Environmental Pattern', may be included in this section. This is quite justified by the realization that under different conditions of the environment the outcome of competition may be different:

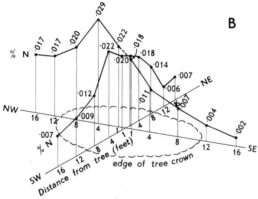

Fig. 8.14. The gradient of pH (*A*) and nitrogen (per cent by weight) (*B*) of surface mineral soil under a tree of *Pinus contorta*. The relative proportions of wind directions in the area are indicated. (From Zinke 1962; courtesy of *Ecology*.)

thus under a deficiency of soil nitrogen species *A* may suppress species *B* in competition although *B* could have grown satisfactorily in the absence of *A*; at a high level of soil nitrogen *B* may well be able to grow with *A* or even suppress it.

Since Watt (1947*a*) outlined several systems in which species were related in cyclical phasic development (see Chapter 4, p. 65), such cyclic phases have been suggested as one of the possible causes of pattern in vegetation. Relatively few further examples of cycles involving several species have been reported in the literature and it would appear that such cycles are rather exceptional and therefore one has to look elsewhere for a general explanation of patterns which are not causally related to environment and morphology. On the other hand the description of cyclical *density* phases of *Pteridium* in a pioneer, building, mature and degenerate series (Watt 1947*a*, 1947*b*) (pp. 69–71), where the performance fluctuates, has been confirmed by several workers since (Chapter 4). This is distinct from cycles in which the actual species change in a recognizable sequence. Watt (1955) shows that *Calluna* follows the same phasic development and furthermore that *Pteridium* as an associated species is inversely related in its distribution to *Calluna*. Thus where *Calluna* is at its building/mature phases, bracken is at its minimum development; conversely where *Calluna* is in its degenerate phase bracken is very much more in evidence. Concurrently, the competitive ability changes with the age series described as pioneer, building, mature and degenerate.

The phasic development of *Pteridium* and *Calluna* have been confirmed by Anderson (1961*a*, and 1961*b*), Nicholson and Robertson (1958) and the 'marginal effect' or 'advancing front' as described by Watt (1947) and representing the pioneer and building phases, has also been reported in several additional species. Thus Ovington (1953), Caldwell (1957) Pénzes (1958, 1960) and Kershaw (1962*a*) have described many examples of advancing fronts of even-aged rhizomes at an optimum level of performance (p. 72). Quantitative data showing the relationship between performance and age in *Alchemilla alpina* and *Carex bigelowii* (Kershaw 1960*b*, 1962*c*) (Figs. 4.12, p. 75 and 4.14, p. 76) together with the general observations on the occurrence of advancing fronts, suggests that the performance of most if not all perennial plants will vary with age. Correspondingly the variation of competitive ability with age (Watt 1955) is likely to be a general phenomenon. Data on the variation of density of *Salix herbacea* as it is invaded by an 'advancing front' of *Eriophorum augustifolium* are given by Kershaw (1962*a*) and confirm the relationship between age and competitive ability. The variation in density of *Salix* taken from a line of contiguous quadrats running through a circular patch of *Eriophorum* actively invading a *Salix herbacea* sward has been discussed previously (p. 73) and the data show the density of *Salix* to be negatively related to the density of *Eriophorum* (Fig. 4.9, p. 73). Furthermore the performance of *Eriophorum* (leaf length) is negatively related to the density of *Salix*. Thus when *Eriophorum* is at an optimum level of performance and density, the density of *Salix* is at a minimum. Austin (1968) in a similar situation has provided evidence of competition between *Zerna erecta* and *Carex flacca*. The internode lengths of the rhizomes of *C. flacca* were markedly shorter when growing within the clumps of *Z. erecta* than elsewhere ($\bar{x} = 1.32 \pm 0.11$ and $\bar{x} = 2.48 \pm 0.18$ respectively, $P < 0.1$ per cent). Since performance varies with age, and the ability to compete varies with the general level of performance of an individual, most if not all perennial plants will have gradually diminishing competitive ability with increase in age. Age as a causal factor of pattern would appear to be of considerable and widespread importance in perennial species, and a mosaic of *density phases* of a species imposed by the varying competitive ability of an associated species is of frequent occurrence and quite independent of any other environmental consideration. The findings of Barclay-Estrup and Gimingham (1969), and Barclay-Estrup (1971), showing strong microclimatic variability induced by the different cyclic phases in *Calluna*, strongly supports this conclusion (see Figs. 4.15, 4.16, etc., p. 79). It also indicates the rather artificial boundary drawn in some instances between 'environmental patterns' and 'sociological patterns'.

A similar pattern effect has been described by Cooper (1960, 1961) who suggests that pattern in ponderosa pine stands results from cyclical development governed by fire. As an old-aged stand degenerates young trees invade and establish, rapidly forming an even-aged group. Such groupings are maintained by the inability of young pines to establish underneath mature parents, and also by periodic fires which remove any young small scattered saplings that do occur. An even-aged stand once established remains as a single unit until the degeneration phase is reached. An area of pine thus consists of a mosaic of phases at different density levels, each group being of a different age.

Sociological pattern is also caused by the effect of an individual on the environment which can be quite independent of the age factor of the individual. The reaction produces a modification of the environment which reduces or enhances the chance of survival of progeny either of the same species or of other species merely, for example, by the annual deposition of leaf litter (see p. 161, Zinke 1962). The interpretation of such instances is by no means as clear cut as would be imagined. The dependence on shrub cover for some herb species can be interpreted as a response to the accumulation of leaf litter or shelter, or protection from grazing, or protection from grass competition, etc., but considerable argument has developed around this phenomenon as to the likelihood of the direct inhibition of associated species by certain shrub species. The case of *Encelia farinosa* is capable of two apparently reasonable interpretations. The lack of ground vegetation under the shrub may be a reflection of toxic substances in its leaves but it may equally reflect the failure of this shrub to accumulate beneath it a fine soil tilth. Other species which contain equally powerful toxins in their leaves and yet can accumulate leaf litter and good soil tilth may also have a dense associated ground flora. The existence in the field of a chemical substance actively exuded by plant roots in quantities sufficiently great to control the establishment or development of associated species, has yet to be demonstrated (see also p. 94), but the possibility of such control of one species by another seems sufficiently probable to necessitate its inclusion in any consideration of pattern.

Fig. 8.15. The arrangement of aerial shoots and rhizomes of *Calamagrostis neglecta* invading bare mud. (From Kershaw 1962; courtesy of *J. Ecol.*)

A very puzzling feature of some rhizomatous species has been reported by Kershaw (1958, 1962) where the aggregation of aerial shoots, arising from different and separate rhizome systems as small 'clumps' 5–10 cm in diameter, is responsible for what at first sight would be regarded as merely a very small-scale morphological pattern.

A typical arrangement of rhizomes and tillers is given in Figs. 8.15 and 8.16 for an area of both recently established and mature *Calamagrostis neglecta* in central Iceland (Kershaw 1962). The pattern is quite marked, the small clumps being composed of at least three, and often four or five different rhizome systems. It has been suggested that the consistent orientation of rhizomes and the production of tillers in the form of a clump,

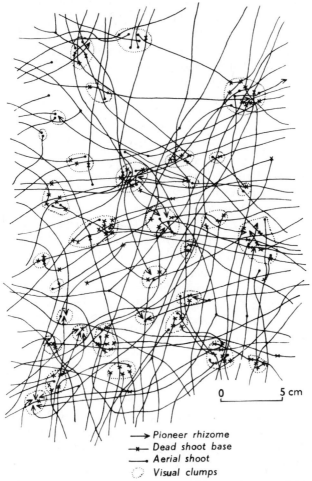

0 —————— 5 cm

→ Pioneer rhizome
—×— Dead shoot base
——• Aerial shoot
⋯ Visual clumps

Fig. 8.16. The arrangement of aerial shoots and rhizomes in an established sward of *Calamagrostis neglecta* in Iceland. (From Kershaw 1962; courtesy of *J. Ecol.*)

may be related to micro-gradients of some environmental factor as yet undetermined, down which the rhizomes grow; finally, tillering occurs at a point where the factor responsible is at its maximum level. However, the possibility does exist that a more orderly and positive mechanism exists. It could be due to the chemical stimulation of a rhizome apex by an established aerial shoot and accordingly this type of pattern is included temporarily at least as a sociological pattern. Whatever the mechanism, it should be emphasized that small-scale pattern should not be immediately dismissed as morphological pattern.

Contagious distribution of individuals (clumping) in vegetation is a common feature, yet little evidence has appeared to suggest that regular distribution of individuals (underdispersion) is of more than rare occurrence in vegetation. The most likely situation which would produce regular distribution is one where there is a high density of individuals within a uniform area; the resulting active competition would limit both the total density over the area, and at the same time the spatial distribution of an individual relative to its neighbours. The lack of data demonstrating regular distribution suggests that either the theoretical conditions leading to competition between individuals for environmental factors are rarely fulfilled, *or are over-shadowed by extreme variations of environmental factors which are able to impose a marked pattern on the species as a whole.* Thus it should be emphasized that even though there is marked contagion at one particular block size, at a similar or larger block size there may be an indication of regular distribution of individuals or of pattern units. As Greig-Smith (1952a) states, 'Professor Bartlett (personal communication) has made a preliminary examination of the problem and his provisional conclusions may be briefly summarized as follows: (1) Variance may be expected to rise to a maximum value as block size reaches the *mean area of clump*, diminishing thereafter if the clumps are underdispersed. . . .' Thus if there is a sharp fall of variance following a marked peak, it can be assumed that the units of pattern have a regular distribution in the area examined.

It is important, however, to bear in mind that such a regular distribution of pattern units is relative to the size of the area investigated, and it is almost certain that an increase in size of the sampling area would show the existence of higher scales of pattern. The evidence to date suggests that frequently there occurs a regular distribution of pattern units up to a limited scale beyond which a further scale of contagion appears. Thus regular distributions of individuals or groups of individuals, rather than being completely absent, are over-shadowed by the scales of contagion present.

In general it is difficult to categorize causalities which are included here under 'sociological' factors for it is virtually impossible to discriminate in the field between effects due to competition and effects due to micro-variations of the environment. The demonstration of positive or negative association between species is straightforward, but determination of competition between species very often necessitates experimentation (see Chapter 3). In certain instances some indication can be obtained from field data: thus Kershaw (1959) showed in *Dactylis* and *Lolium* a sociological scale of pattern at 80 cm. His decision was based on a significant negative association at high densities and nil association or even a slight positive association at low densities. This change of trend in association with increasing density does suggest competition rather than effects of slightly dissimilar environmental requirements.

A similar example is given by Vasilevich (1961) for *Cladonia sylvatica* and *C. rangiferina* in the ground vegetation of pine forest in Russia. The two species are positively associated at low densities and negatively at high densities, again suggesting competition between the species at high densities. A further example of sociological pattern which shows negative association between species which is, in part at least, related to competition is taken from the relationship between *Carex bigelowii* and *Festuca rubra* (Kershaw 1962c) (p. 143). Both the density and performance data of *Carex* are positively related to micro-topography and negatively related to the density of *Festuca*. However, the performance of *Carex* is more markedly affected by *Festuca* than is the density of *Carex*, which again suggests an effect of competition. (See also Fig. 7.9, p. 144.)

GENERAL CONSIDERATIONS

The widespread detection of non-randomness in vegetation, usually in the form of contagion or clustering of individuals, has led to the realization that all vegetation will contain one or many scales of pattern. The only possible exception to this is in the colonization of a homogenous bare area when the initial migrules may be distributed more or less at random. Morphological scales of pattern will always be present, and it is rare to find an area with such a completely uniform environment as not to produce one or more scales of environmental pattern. The presence of pattern in environmentally uniform areas is also to be expected since intrinsic properties of the plants themselves will interact to produce pattern. Thus it is difficult to envisage an area where all the individuals are of an even age and hence have equal levels of performance and competitive ability. Accordingly we are forced to the conclusion that pattern will always be present in vegetation except in a few rare and extreme cases.

9 The Detection of Natural Groupings of Species: Classification Methods

THE two opposing schools of thought who on the one hand consider the climax association to be a complex organism or quasi-organism and on the other hand those who support the individualistic concept, also reflect the present-day ecological approaches to the delimitation and definition of community units. Thus the concept of the complex organism implies considerable interaction between the species which jointly modify the environment and form a distinguishable vegetational group. Conversely the individualistic concept regards no two communities as being identical, but considers that they show a continuous variation and accordingly can not be readily delimited as definable units. Numerous attempts have been made in recent years to extract statistically or objectively definable units from vegetation, by means which range from complex statistical procedures to rather simple subjective methods. On the other hand there has been a growing awareness that classification of units was in many instances an artificial step. Intermediates were being ignored and this variation, very often of continuous form, was important. Accordingly methods of *arranging* samples in order of similarity appeared in the literature, and recently the application of multi-variate techniques to this field has resulted in a bewildering array of coefficients, techniques and terminology. An attempt will be made to cover this area in a following chapter.

Of fundamental importance is the difference of purpose between classification of plant units or arrangement—*ordination*—of samples. The nature of the unit and its spectrum of species is the main aim of classification. Conversely, the continuous variation expressed in an ordination implies environmental control and the interest is centred around the environmental parameters concerned in the ordination.

THE BRAUN-BLANQUET SCHOOL OF PLANT SOCIOLOGY

The aim of the Braun-Blanquet school (Braun-Blanquet 1932, 1951) is a world wide classification of plant communities based on a number of concepts which have been critically reviewed by Poore (1955,

1956). The initial attitude of the British ecologists to these methods and to the results was one of extreme criticism, but more recently attempts have been made to understand more fully what the Continental workers are attempting to achieve and the vegetation of the Scottish Highlands has been comprehensively treated by McVean and Ratcliffe (1962) on this basis.

The foundation of the classificatory system is the *association* which is the basic vegetational unit and is an abstraction obtained from a number of lists of species from selected sites in the field.

Table 9.1 gives an association table for the *Rhacomitrium/Carex bigelowii* association made up of twelve lists based on the Domin scale with details of each stand incorporated at the head of the table. From such an association table it is possible to pick out those species which occur in a given number of stands, i.e. the *degree of presence* can be assessed for all species. Braun-Blanquet expresses the degree of presence of a species in a five-degree scale:

5 = constantly present (in 80–100 per cent of the stands).
4 = mostly present (in 60–80 per cent of the stands).
3 = often present (in 40–60 per cent of the stands).
2 = seldom present (in 20–40 per cent of the stands).
1 = rare (in 1–20 per cent of the stands).

Thus in the *Rhacomitrium/Carex* association table (Table 9.1) *Carex bigelowii, Dicranum fuscescens, Polytrichum alpinum, Rhacomitrium lanuginosum, Cetraria islandica* and *Cladonia uncialis* occur with a degree of presence 5; similarly *Vaccinium myrtillus* and *Festuca vivipara* have a degree of presence 4, and so on down to such species as *Lycopodium selago, Nardia scalaris, Coriscium viride*, with a degree of presence 1. Where the data of species presence is derived from samples taken from plots of known and limited area the degree of presence is referred to as the *degree of constancy*.

CONSTANCY

It is clear that if a very large sample area is used many more species will be 'constant'; this leads to a discussion of the minimal area concept which will be postponed to a later section (see below). The method of obtaining the lists and measures used have been described fully in Chapter 1, but several further concepts are of considerable and fundamental importance.

CHOICE OF SITE

The association tables are derived from a large number of *carefully chosen sites* and it is clear that the whole basis of the system is established

Table 9.1   *Rhacomitrium*/*Carex bigelowii* association table. (From Poore 1955; courtesy of *J. Ecol.*)

No. of sample plot	520010	520011	520035	520064	520068	520076	520117	520162	520167	520144	520006	520193
Empetrum hermaphroditum	2	–	–	–	4	1	–	–	–	–	–	–
Salix herbacea	–	–	–	–	–	–	3	–	–	–	–	–
Vaccinium myrtillus	5	3	–	–	4	3	3	3	3	4	4	–
V. vitis-idaea	–	1	–	–	2	2	2	–	1	–	–	–
Lycopodium selago	3	–	–	–	–	2	–	–	–	–	–	–
Agrostis canina	–	3	3	–	–	3	–	–	–	–	–	–
A. tenuis	1	–	–	–	2	–	–	–	–	–	–	–
Deschampsia flexuosa	5	–	–	–	–	3	3	–	–	–	–	–
Festuca ovina agg.	3	–	–	–	–	–	–	–	–	–	–	–
F. vivipara	–	–	–	3	3	4	2	3	1	2	2	1
Nardus stricta	1	–	–	–	–	–	–	–	–	–	–	–
Carex bigelowii	4	4	4	4	3	3	3	5	4	4	6	9
Luzula multiflora	–	–	–	–	–	–	–	–	1	–	–	1
L. spicata	–	–	–	–	–	1	2	–	–	–	–	–
Alchemilla alpina	–	–	–	–	–	3	1	–	–	–	2	–
Galium hercynicum	–	–	2	–	–	2	2	–	–	–	2	1
Dicranum fuscescens	1	1	2	2	2	–	3	1	4	3	3	3
D. scoparium	–	3	–	–	–	1	1	2	–	–	–	–
Aligotrichum hercynicum	1	3	–	–	–	–	1	–	–	–	–	1
Pleurozium schreberi	–	–	–	–	–	1	1	–	–	–	4	–
Pohlia? annotina	–	1	–	–	–	–	–	–	–	–	–	–
Polytrichum alpinum	3	–	1	3	2	2	3	3	1	3	3	2
Rhacomitrium lanuginosum	9	7	10	10	9	9	10	10	10	10	6	9
Rhytidiadelphus loreus	–	–	1	–	–	–	–	–	–	–	1	–
?Aplozia sphaerocarpa	1	–	–	–	–	1	–	–	–	–	–	–
Diplophyllum albicans	1	–	–	–	1	1	–	–	–	–	–	–

Species	1	2	3	4	5	6	7	8	9	10	11	12
Nardia scalaris	—	—	—	—	—	1	—	—	—	—	—	—
Lophozia? ventricosa	—	—	—	—	—	1	—	—	—	—	—	—
Ptilidium ciliare	—	—	—	—	—	—	—	—	—	—	—	—
Alectoria nigricans	—	1	1	1	1	1	1	1	2	1	1	—
Cerania vermicularis	1	1	—	1	2	3	2	1	2	2	2	1
Cetraria aculeata	2	1	1	1	1	2	1	1	2	2	3	1
Cetraria islandica	1	2	3	2	—	—	2	2	3	3	2	3
Cladonia alpicola	1	2	—	1	—	—	1	1	—	—	1	1
C. bellidiflora	1	—	1	1	1	—	1	1	—	—	1	—
C. coccifera	—	—	1	1	1	—	—	—	—	—	—	—
C. crispata	—	—	—	1	—	—	—	—	—	—	—	—
C. furcata	—	1	1	1	1	2	—	1	2	—	1	2
C. gracilis	1	—	—	1	1	—	1	—	—	1	1	1
C. pyxidata	—	—	1	—	—	—	1	1	1	—	1	—
C. rangiferina	—	1	1	2	2	—	2	1	—	2	1	2
C. sylvatica	—	—	2	1	3	2	3	3	2	2	1	1
C. squamosa	1	1	1	—	1	1	1	1	1	1	1	—
C. uncialis	4	3	2	3	3	3	3	3	3	3	3	1
Coriscium viride	1	1	—	1	1	1	—	1	—	—	—	—
Icmadophila sp.	—	—	—	—	—	—	—	—	—	1	—	—
Ochrolechia frigida	—	—	1	—	1	—	2	1	—	1	1	—
Sphaerophorus globosus	—	—	1	—	2	1	1	1	—	—	1	—

on a markedly non-random sampling procedure. The first important proviso in the choice of a sample plot is its homogeneity. Dahl and Hadač (1949) define homogeneity in the following way:

> A plant species is said to be homogeneously distributed in a certain area if the possibility to catch an individual of a plant species within the test area of given size is the same in all parts of the area. A plant community is said to be homogeneous if the individuals of the plant species which we use for the characterization of the community are homogeneously distributed.

The accumulated evidence relating to the species distribution patterns present in all vegetation types so far examined (Chapter 8), throws considerable doubt on the existence of the homogeneous plot ideally suitable for formulation of the association table. Dahl and Hadač recognize the difficulty and state:

> 'In nature plant communities are never fully homogeneous, i.e. the density of the species of importance by the characterization of the community, especially the dominating species, have not exactly the same density in all parts of the area. We may thus talk of more or less homogeneous plant communities. Measurements on the homogeneity of the plant community are rarely carried out, but the human eye, badly adapted to measurement but well to comparison, rapidly gives the trained sociologist an impression whether a plant community he has before his eyes is highly homogeneous or not.'

Again this seems extremely doubtful since a large number of patterns have been detected in areas which appeared absolutely uniform even to the most careful inspection. This aspect of the method has also been severely criticized by Goodall (1954) who states '. . . and it is pointed out that no area of vegetation has ever been shown to be fully homogeneous. It is even doubted whether areas with greater and smaller variance between replicate samples may be distinguished in natural vegetation.'

MINIMAL AREA AND HOMOGENEITY

Assuming there is a reasonable degree of uniformity the second proviso is that the stand must be sufficiently large to enable a representative sample to be taken. This minimum size requirement of the sample is termed the *minimal area* and is related to the number of species which occur as the sample plot increases in size. With a small plot the number of species present is also small but as the size of the plot is increased the number of species increases very noticeably. Eventually the number of new species found in successively larger plots becomes progressively fewer.

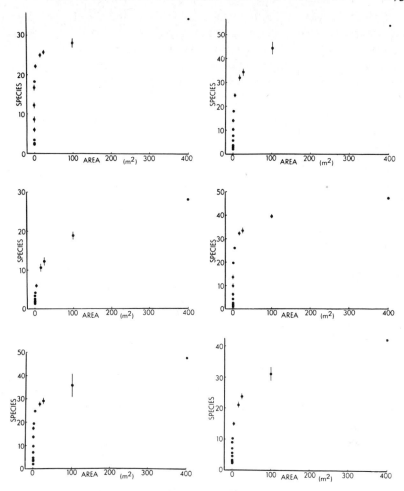

Fig. 9.1. Species area curves from six sample sites. (From Hopkins 1957; courtesy of *J. Ecol.*)

This general trend is apparent in a graph of the number of species against the area of the sample plot, or a species area curve as it is usually called. The Braun-Blanquet concept of minimal area is related to the area at which the species-area curve becomes approximately horizontal. A number of examples of species-area curves are given below (Fig. 9.1) all broadly similar and of characteristic form. It is apparent that though the slope of the curve changes fairly markedly there is a continued increase in the number of species as the area of the sample plot is further enlarged. Hopkins (1957) presents a clear account of the use of species-area curves relative to the determination of minimal area (see also Goodall 1952,

Greig-Smith 1957) and draws attention to the extreme difficulty of establishing any certain definition of the minimal area. Thus the concept of minimal area depends on a definite 'break' in the curve. Cain (1932, 1934) claimed that such a break could be 'recognized' but later (1938) pointed out that its position depended on the ratio of the axes used. Following this

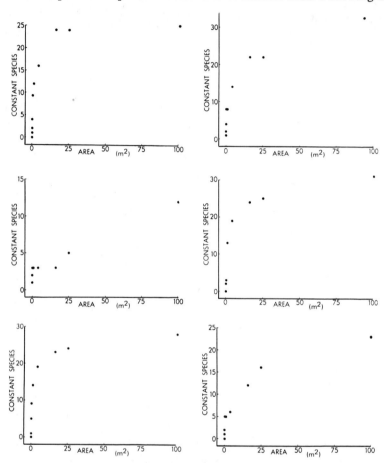

Fig. 9.2. Constancy-area curves from six sample sites (From Hopkins 1957; courtesy of *J. Ecol.*)

a number of attempts to determine the minimal area based on the ratios of species and area have been made. A similar approach was made by DuRietz (1921) who related the number of constant species to area in constancy-area curves, 'constants' being defined as species which have a percentage frequency greater than 90 per cent and the minimal area defined as 'the smallest area on which an association attains a definite number of constants'. Hopkins (1955) gives a number of constancy-area curves (Fig. 9.2) of a very similar form to the species-area curves and with the same attendant difficulty in determining the position of the minimal area. He concludes: 'Thus minimal area cannot be determined objectively from the species-area or constancy-area curve.'

The problems of homogeneity and minimal area outlined above are inter-related. If true homogeneity existed then minimal area would be a valid concept. However it is clear from the existence of pattern at numerous scales in apparently homogenous vegetation that minimal area can never be more than a gross approximation and accordingly some subjective judgement is necessary in assessing whether the area sampled is large enough to reflect the characteristics of that particular community.

A more recent review of the Braun-Blanquet system by Moore (1962) avoids the problems of homogeneity and minimal area by accepting the subjectivity of the approach and suggesting that a long 'tail' of isolated occurrences at the foot of an association table is a sign of a badly chosen stand. The fact that this does occur is clear evidence of the impossibility of visually assessing homogeneity and the minimal area required for sampling even by an experienced worker.

THE ASSOCIATION

The mechanical construction of the association table has been described above and the subsequent steps are a careful examination of the tables and separation of any lists which differ consistently and noticeably from the rest. These differences vary considerably from minor variations to more obvious differences which may necessitate the establishment of another association. Poore (1955) sums up this stage of the procedure as 'muddled and haphazard'. 'According to Braun-Blanquet these decisions can only be made by having recourse to a consideration of the fidelity of the various species, and this stage (that of distinguishing associations by faithful species) is one about which the literature is particularly reticent. Apart from the proviso that it requires much time and experience we are told nothing, and observation of the system in use has done nothing to unravel the problem for me.' This seems to be a fair comment on the abstraction of the associations that have been established. Fidelity is a characteristic of certain species which shows a degree of exclusiveness towards a particular association. Thus Braun-Blanquet (1952) lists five degrees of fidelity.

FIDELITY 5. Exclusive species, completely or almost completely confined to one community.

FIDELITY 4. Selective species, found most frequently in a certain community but also rarely, in other communities.

FIDELITY 3. Preferential species, present in several communities more or less abundantly but predominantly in one certain community and there with a greater degree of vigour.

FIDELITY 2. Indifferent species, without a definite affinity for any particular community.

FIDELITY 1. Accidentals, species that are rare and accidental intruders from another community or relics of a preceding community.

Poore is correct in saying the only situation in which degrees of fidelity can be properly assessed is that which exists when all the vegetation of a region has been described. He quite rightly lays emphasis on 'constancy' to delimit the association groupings rather than determining the association by the species faithful to them (fidelity grades 3–5). The difference between 'constancy' and 'fidelity' may not at first sight be clear but would seem to be related to the ecological tolerance of the species. Thus a 'constant' species may have a wide ecological tolerance and occur in several associations, while a species with a high degree of 'fidelity' has a narrow ecological tolerance and only occurs in one association. The situation is rendered even more confusing by an example quoted by Poore (1955a) taken from Barkman (1940). The example is an epiphytic association and the association table comprises fourteen lists with seven faithful species. Of the fourteen lists, one contains 5 of the faithful species, one 3, two 2, eight 1, and two 0. Poore expresses his difficulty in deciding which criteria were adopted in retaining the majority of these lists in this association. It would appear that either the associations are based in fact on constancy, in which case degree of fidelity seems unnecessary, or the sites are chosen because species of known narrow ecological amplitude are present, and these species subsequently, by circular argument, are defined as the faithful species. Thus faithful species are those which are confined to one association and the association is recognized by these faithful species; this is perfectly legitimate if the association is delimited by other characteristics in the first place. If in fact the association is recognized by its faithful species it is then not valid to define the faithful species from the associations. Moore (1962) denies the over importance of fidelity and quotes Braun-Blanquet (1959): 'It should not be concluded, as has sometimes been maintained, that the associations are based on fidelity. This is emphatically not the case. The association is an abstraction based on the totality of more or less homogeneous relevés (lists) which floristically correspond closely to one another. It is however, characterized, not merely floristically but also ecologically, dynamically and geographically. Nevertheless, in distinguishing from one another the associations which have been conceived on a floristic basis, far greater importance and more general significance is attributed to fidelity than to purely quantitative characteristics especially when fidelity is combined with high constancy.' This latter instance would seem to be of importance and considerable value, but as has been pointed out only possible when the complete vegetation of a region has been described. Accordingly Poore's approach based on constancy seems fully justified.

However, it is still a subjective decision to separate out a number of lists from an association table which have one or two species in common and set up a further association. There seems to be no objective reason why the splitting should stop there. Moore (1962) attempts to elucidate this point but states: 'In order to understand this (how the association under discussion is arrived at), it is essential to grasp the principle of the differential species.' It is this tendency for emphasis of fidelity and the suspicion of a circular argument which so reduces the value of the approach. Poore's insistence on the use of the 'constant' species to delimit the 'association' is a great improvement but subjective decisions still have to be taken as to the 'level' of the association and whether a certain group of lists legitimately represent a fragment of a further association. Poore (1956) used a method of successive approximation; if no obvious constants were present in the lists they were taken together and divided on probable differential species (species consistently present in one group of lists). Provisional constants were named and sites were sought which possessed these combinations of constants. If the constants were real such stands were readily found but if not, they proved impossible to find.

Braun-Blanquet, having established a number of associations then classifies them into a hierarchy, which is a convenient if an unnatural system. The shortcomings of the continental system are rather numerous and have been discussed above under each step of the procedure. The modifications as proposed by Poore (1955, 1956) to a large extent remove some of the major criticisms and a workable system is available for the rapid description and characterization of a large area of vegetation. However, the sampling procedure, i.e. the choice of a sample site, is entirely subjective, and the apparent sharpness of the limits of associations are in fact a result of this non-random method of sampling. The sites which are clearly intermediate are avoided as 'not homogenous' and as large numbers of intermediates exist, even if they occur rarely, the sociological approach involving avoidance of such sites over-simplifies the picture to a considerable extent. This by no means invalidates the real existence of the associations that have been erected and they are obviously of frequent and widespread occurrence. A more real interpretation of the association should be based not only on its floristic composition but also on its frequency of occurrence in a region. The term 'noda' introduced by Poore is a satisfactory one and does imply the existence of intermediates of less frequent occurrence relative to the 'central type'. The difficulty of finding an area which is homogenous and of accurately defining the minimal area relative to it is unavoidable. It is worthwhile to accept these limitations within the system outlined by Poore, since it can readily contribute considerably to our knowledge of the vegetation especially in those regions where we have little or no floristic information at all.

OBJECTIVE METHODS

The subjective nature of the delimitation of vegetational units as proposed by the Continental and Scandinavian sociologists have led to numerous attempts at establishing similar units by objective methods. If we examine two noda which are positively related to some extent, in both their species composition and also their environmental relationships, we can consider them as an abstraction in terms of their species and also in terms of their relationships with *other* noda or plant associations.

In fact this situation has several important mathematical properties, but of immediate relevance is that species A–G and species I–N are positively associated together, but group A–G is negatively associated with group I–N. One further important property of this simple model is that only 'characteristic species' are being utilized. Species H and species O–Z may be present or not but do not contribute any significant difference between the two groups. Statistically then:

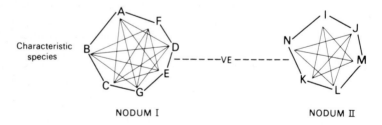

Objective methods of classification utilize and optimize this situation and several conveniently use χ^2 as a measure of positive and negative association. Two techniques will be considered illustrative of this approach.

ASSOCIATION ANALYSIS

The initial contribution to these methods was made by Goodall (1953). It is of considerable interest and will accordingly be described in some detail. 2 × 2 contingency tables were prepared from presence and absence data of all species in a *Eucalyptus oleosa/E. dumosa* association in Australia. The area was topographically uneven with sandy ridges running east and west separated by hollows with a heavier soil. An area 640 m square was delimited and divided into 64 square plots within each of which four 5 m quadrats were placed at random. Fifty-eight species were recorded from the area but 27 only occurred in less than 5 quadrats and were eliminated. χ^2 values were calculated for the occurrence of the remaining species in pairs. Only positive associations (in the sense that the species were positively correlated) were used in the classification of the quadrat data. Goodall points out that homogenous groups of quadrats can be obtained (which is the object of a system of classification) by eliminating correlation between species in either of four ways.

(*a*) Quadrats with one of two species which are correlated can be separated out as an initial potentially homogeneous group.

(*b*) Quadrats not containing one of the species represent a similar potentially homogeneous group.

(*c*) Quadrats containing both the species.

(*d*) Quadrats containing neither of the species.

Goodall analysed the data by the four methods and found reasonable agreement in the results, and concluded that the first system, the presence of a pair of correlated species, was the most convenient method of approach.

The first step was to separate out from the total sample (256) those quadrats containing *Eucalyptus dumosa* which showed a number of highly significant correlations with other species,* and was the most frequently occurring species in the area (170 quadrats). Within this group of quadrats several highly significant positive correlations were still present and a second division was made, again on the most frequent species, *Triodia irritans*, which showed a highly positive correlation with some other species. One hundred and nineteen quadrats were separated out and again fresh contingency tables established and the heterogeneity of these 119 quadrats examined. The process was continued with successively smaller groups of quadrats until finally 53 quadrats selected on the presence of *Eucalyptus dumosa*, *Triodia irritans* and *Stipa variabilis* showed no significant correlation between species within the group and

* This in fact is 'association' between species that is measured but the term correlation is employed here to avoid confusion with the use of 'association' in the Braun-Blanquet methodology.

were finalized as a homogenous group (group *A*). All the remaining quadrats (203 in all) were pooled and a fresh series of contingency tables constructed. The same procedure was followed and the most frequent species *Eucalyptus oleosa*, in this residual group of quadrats was used as the initial division (121 quadrats). Subsequently 92 quadrats containing *Bassia uniflora* and finally 17 quadrats containing *Cassia eremophila* were separated out.

Table 9.2   Frequency (per cent) of species in four groups of quadrats based on presence of single species showing positive correlations. (From Goodall 1953; courtesy of *Austr. J. Bot.*)
Frequencies of 50 per cent or more have been given in bold type

Species	Group			
	A + E	*B*	*C*	*D*
	Number of quadrats			
	83	17	66	90
Acacia rigens A. Cunn.	2	0	5	0
Bassia parviflora Anders	2	0	0	4
B. uniflora R. Br.	17	**100**	5	**100**
Beyeria opaca F. Muell.	1	6	5	0
Callitris verrucosa R. Br.	1	0	26	0
Cassia eremophila A. Cunn.	1	**100**	0	2
Chenopodium pseudomicrophyllum (1)	5	6	2	22
C. pseudomicrophyllum (2)	12	6	0	13
Danthonia semiannularis R. Br.	20	0	6	7
Dodonaea bursariifolia Behr.	**53**	0	**71**	17
D. stenozyga F. Muell.	1	0	0	6
Enchylaena tomentosa R. Br.	4	24	0	3
Eucalyptus calycogona Turcz.	14	18	11	30
E. dumosa A. Cunn.	**81**	12	**100**	39
E. oleosa F. Muell.	49	**100**	24	**81**
Grevillea huegelii Meiss	5	35	8	4
Kochia pentatropis R. Tate	2	24	0	7
Lepidosperma viscidum R. Br.	1	0	9	0
Lomandra leucocelphala R. Br.	4	0	8	0
Melaleuca pubescens Schau.	4	0	8	0
M. ucinata R. Br.	**55**	0	**53**	3
Micromyrtus ciliatus J. M. Black	0	0	9	0
Myoporum platycarpum R. Br.	2	0	5	0
Olearia muelleri Bth.	5	**59**	0	6
Santalum murrayanum F. Muell.	1	12	5	1
Stipa elegantissima Lab.	1	12	0	3
S. mollis R. Br.	7	0	5	10
S. variabilis Hughes	**87**	**88**	0	**78**
Triodia irritans R. Br.	**78**	6	**100**	3
Vittadinia triloba D.C.	46	24	11	37
Westringia rigida R. Br.	6	**71**	5	3
Zygophyllum apiculatum F. Muell.	4	**65**	2	47

This final group of 17 quadrats showed no significant correlations and were established as group B. The remaining 186 quadrats were again examined by 2×2 contingency tables, divisions being based either on the most frequent species which showed significant correlation or on the next

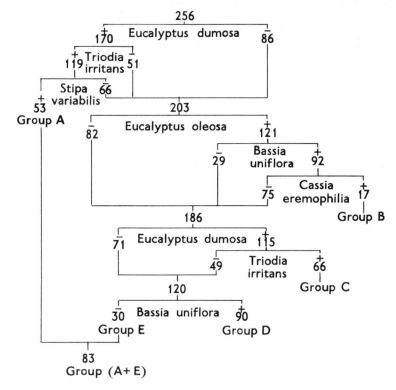

Fig. 9.3. Victoria Mallee vegetation. Stages in the classification of quadrats into groups. At each stage the number of quadrats in which the species in question is present ($+$) and absent ($-$) is indicated. (From Goodall 1953; courtesy of *Austr. J. Bot.*)

most frequent species with a significant correlation. After the final separation of a homogenous group the total remaining quadrats were subsequently examined each time. The data was divided by this means into five homogenous groups assessed by χ^2 tests. Each group was finally recombined with each of the other groups in turn and tested for heterogeneity. If the homogenous groups were satisfactory then recombination with other groups would immediately introduce correlation. This was found to be so, with the exception of groups A and E which on combination showed no significant correlations and were accordingly combined as a single unit (Fig. 9.3). The species components of the groupings are given in Table 9.2. Goodall characterizes the four groups *A*, *B*, *C* and *D* by *Vittandinia triloba, Cassia eremophila, Dodonaea bursariifolia* and *Bassia uniflora* based on those species of most frequent occurrence in those quadrats allotted to the same grouping by all four procedures. The distribution of the groupings in relation to the area studied (Fig. 9.4) shows

○ *Dodonaea bursariifolia* grouping

× *Vittadinia triloba* grouping

△ *Bassia uniflora* grouping

□ *Cassia eremophila* grouping

Fig. 9.4. Plan of the area of Mallee vegetation studies indicating the position of each quadrat and the grouping to which it was assigned. (From Goodall 1953; courtesy of *Austr. J. Bot.*)

a reasonable correlation with topography. Three ridges run across the area and are occupied predominantly by the *Dodonaea* and *Vittandinia* groupings while the hollows between contain the *Bassia* and *Cassia* groupings.

A similar approach leading to a hierarchical classification has been outlined by Williams and Lambert (1959, 1960). They have three main criticisms of Goodall's method: his use of positive correlations only, his use of the most frequent species as a basis of division of a potentially homogenous group, and the pooling of those quadrats which do not form the initial and successive homogenous groups. The use of negative correlations as well as positive ones greatly strengthens the analysis, as long as due regard is given to the size of quadrat used for sampling since negative correlation between species may frequently occur by virtue of the fact that no two individuals can occupy the same place at any one time. Their criticism of the 'pooling' of quadrats seems equally logical: 'Imagine a population divided on species X into (X) and (x). (X) is still found to be

Fig. 9.5. The Shatterford community analysis. (From Williams and Lambert 1960; courtesy of *J. Ecol.*)

divisible on species Y into (XY) and (Xy). Goodall would "pool" (Xy) with (x) to form a new population. Our chief objection to this system is that information relating to the discontinuity $(X)/(x)$ is being discarded, and on these grounds we would prefer a "hierarchical" system—a division, once made, would remain inviolate throughout the analysis.' It is also

Fig. 9.6. The Shatterford community map, soil type, and contours. (From Williams and Lambert 1960; courtesy of *J. Ecol.*)

pointed out that Goodall's final groups are not necessarily capable of definition in terms of key species and also that the means by which the groups are obtained do not give any meaningful information.

2×2 contingency tables are used to establish a χ^2 matrix and the χ^2 values are summed irrespective of whether they reflect a positive or a negative correlation. The species with the largest $S(\chi^2)$ and not the most abundant species is used in the sub-division of the sample, the two quadrat groups thus formed then remaining separate throughout the remainder of the sorting procedure. Each of the two groups is re-examined and further divided on the species with the largest $S(\chi^2)$. The hierarchical division is continued until no significant correlations appear, at which point the group is designated as 'final'. Williams and Lambert give several convincing examples of the application of the method to actual field data. The Shatterford example involving 29 species is extremely clear-cut, the analysis (Fig. 9.5) separating out six major groups which are further sub-divided into a larger number of smaller units. The spatial distribution of the individual quadrats in relation to the soil variations of the area is remarkably consistent (Fig. 9.6). The first division (community 4) represents the area of dry heath dominated by *Calluna vulgaris* with *Erica tetralix* and *Molinia caerulea*. Communities 5 and 6 are typical bog communities dominated by *Erica tetralix* and *Molinia caerulea* respectively with community 1 occupying an intermediate 'wet heath' position (Table 9.3). From the species composition of the groupings and the relative frequency of occurrence the 'constant' species of the groupings are readily apparent. Thus the analysis appears to accomplish all that was intended for it and has the great advantage (in common with Goodall's approach) that the position of each group of quadrats can be accurately plotted on a map and related to the detailed environmental variations of the area as a whole. Thus any group of quadrats can be picked out at a later date and the environmental inter-relationships considered in detail.

In terms of our simple model:

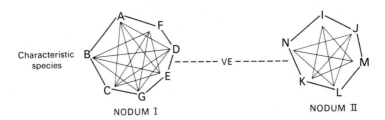

Goodall's approach simply uses the $+^{ve}$ relationships existing within the two groups of species. Williams on the other hand utilizes in addition, the $-^{ve}$ relationships existing *between* the two groups. Both methods appear

Table 9.3   Species composition of Shatterford groupings at six-community level. (From Williams and Lambert 1960; courtesy of *J. Ecol.*)

Ref. number of community	4	2	3	1	5	6
Number of quadrats	214	26	17	72	50	125
Erica cinerea L.	29	19	–	–	–	–
Ulex minor Roth	22	22	–	–	–	–
Festuca ovina L.	–	26	–	–	–	–
Pinus sylvestris L.	1	–	1	–	–	–
Calluna vulgaris (L.) Hull	214	26	17	72	–	–
Polygala serpyllifolia Hose	8	19	5	30	1	–
Erica tetralix L.	155	26	17	72	50	–
Molinia caerulea (L.) Moench	155	26	17	72	50	125
Trichophorum caespitosum (L.) Hartman	–	3	17	70	10	–
Drosera rotundifolia L.	–	–	7	58	23	4
Juncus squarrosus L.	–	–	–	5	–	–
Gentiana pneumonanthe L.	–	–	–	2	–	–
Eriophorum angustifolium Honck.	–	–	–	72	50	77
Narthecium ossifragum (L.) Huds.	–	–	–	57	49	25
Drosera intermedia Drev. and Heyne	–	–	–	9	15	–
Carex panicea L.	–	–	–	1	32	4
Potentilla erecta (L.) Rausch.	–	–	–	1	2	2
Myrica gale L.	–	–	–	4	47	76
Juncus bulbosus L.	–	–	–	–	1	2
Carex rostrata Stokes	–	–	–	–	1	8
Juncus acutiflorus Hoffm.	–	–	–	–	4	37
Potamogeton polygonifolius Pourr.	–	–	–	–	2	23
Agrostis ? gigantea Roth	–	–	–	–	1	20
Carex echinata Murr.	–	–	–	–	1	28
Scutellaria minor L.	–	–	–	–	–	26
Cirsium dissectum (L.) Hill	–	–	–	–	–	15
Hypericum elodes L.	–	–	–	–	–	3
Potamogeton natans L.	–	–	–	–	–	2
Eleocharis pauciflora (Lightf.) Link	–	–	–	–	–	1

to work satisfactorily and associate species together into groupings which in turn relate to the ecology of the area sampled. The methods, however, both follow a different pathway of sub-division and furthermore the divisions are based on the commonest species in one method and the species with the strongest level of association in the other. Additional methods of sorting out 'groupings' in objective ways are largely concerned with optimizing the efficiency of classification, and two of the evaluation criteria used are in fact the pathway followed and the method of sub-division. The major factor for assessment of the techniques, however, is the credibility of the result. A considerable body of literature has built up relating to objective methods of grouping and classifying data, and details are available in Williams *et al.* (1966), Williams and Dale (1965) and Sokal and Sneath (1963).

In an ecological context, variation between samples is continuous and there is no implicit evolutionary relationship between what noda are present. Conversely in a taxonomic situation discontinuities *are* usually present and one objective of classification is the uncovering of possible phylogenetic relationships. In the sense of continuous variation, classification of vegetational samples is thus an artificial concept which is achieved by either ignoring intermediates in the sampling or conversely by ignoring that determinate characteristic of a nodum, the relative abundance of its component species. Goodall and Williams both achieve groupings by utilizing presence–absence data and the resultant clarity is simply due to this. Thus in ecology, in a situation of continuous variation, there is much to be said for simplicity of approach if an arbitrary classification is required; the complexities of methodology should not be allowed to dominate. The use of χ^2 as a measure of association suffers the disadvantage that the trend of association can be affected by the size of quadrat

Fig. 9.7. The inverse association analysis of the Shatterford data. (From Williams and Lambert 1961; courtesy of *J. Ecol.*)

used in the sampling. Kershaw (1961) has shown that the final quadrat groupings in association analysis depend largely on the size of the quadrat used. The primary divisions are not affected, but such overall differences are usually apparent on visual inspection of the area and do not necessarily need an elaborate technique to pick them out. On the other hand these methods of association analysis are objective, and for the primary survey of an area very effective. The initial problem of computing large numbers of χ^2 values has been largely removed by the subsequent development of computers with very large central stores and/or peripheral backing store. The original limitations on the size of sample and number of species which could be handled in any one analysis have now disappeared to a great extent.

The approach of Williams and Lambert (1959, 1960) is more efficient than that of Goodall (1953) in that negative associations are also included in the analysis, and despite the short comings of χ^2 as a measure, its relative simplicity recommends it. One further aspect is of importance. Williams and Lambert (1961) showed that in addition to the 'normal' analysis of plot data, a similar but 'inverse' analysis could be made of the species inter-relationships.

For species A, B, C, D in sample plots 1, 2, 3, 4, calculation of the number of joint occurrences for entry into 2×2 contingency tables of the 'normal' or plot analysis are calculated as follows:

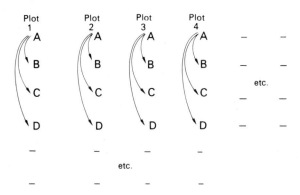

Species $A+B$, $A+C$, $A+D$, etc., in this case, each show four joint occurrences in the plots. From species comparisons, the species with the strongest level of association is selected and the *plots* sub-divided on this species.

Conversely in an inverse analysis:

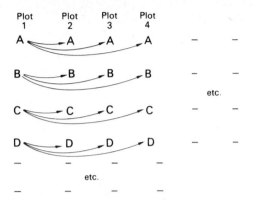

Plots $1+2$, $1+3$, $1+4$, etc., in this case, each show four joint occurrences of species.

From plot comparisons the plot with the strongest level of association is selected, and *species* sub-divided in relation to this plot. Thus those species present in that plot form a group of species which are subsequently re-examined throughout the sample for further possible sub-division. The final outcome is species groupings whose ecological preferences are closely similar.

The results from an inverse analysis of the Shatterford data previously discussed are given below (Fig. 9.7) and show four species groupings. Group A comprised of *Calluna vulgaris*, *Molinia caerulea*, and *Erica tetralix*, a very characteristic combination of wet heathland species, Group B with *Trichophorum caespitosum*, *Eriophorum angustifolium* and *Narthecium ossifragum* typical of wet heath/bog, Group C composed of the dry heath species *Erica cinerea*, *Ulex minor*, *Festuca ovina*, and *Polygala serpyllifolia*. The final group D is a group of species of wet boggy habitats, but whose representation in the sample is poor. The 'inverse analysis' thus complements the 'normal' association analysis findings.

There has been considerable confusion of terminology following the application of the more recent multi-variate techniques to the problem of vegetational groupings. The so-called R & Q techniques, despite an attempt by Williams and Dale (1965) to clarify the situation, are still utterly confusing to the non-mathematician. Accordingly it is advisable, and much simpler, to use the terms *plot classification* and *species classification* which are readily understandable, together with sufficient detail of the actual method used so as to avoid further confusion. This nomenclature will be followed here and also extended to plot and species ordinations subsequently (see below).

The simplicity and effectiveness of *plot* and *species* classifications probably accounts for their widespread use. If classification seems a useful approach then the over-simplification inherent in the χ^2-methods of analysis is probably beneficial in the first instance. Once knowledge has accumulated as to the composition of the 'central types' of associations, the peripheral variants and intermediates can be subsequently included using the more powerful but more complex 'clustering techniques' now available (see below). Plot and species classifications on a range of temperate vegetation types are given by Gittins (1965), Gimingham *et al.* (1966), Harrison (1970) and Ward (1970). Similarly Flenley (1969), Hopkins (1968), Poore (1968), Greig-Smith *et al.* (1967), and Kershaw (1968), have all applied the techniques to tropical vegetation. Many of the examples cited above also compare the approach with ordination methods and show a reasonable degree of agreement between 'clusters' arrived at by plot ordination and the results obtained by plot classification. Despite the emphasis of presence and absence criteria in the method, as opposed to abundance measures which give a more realistic account of a plant association, association analysis has given fairly reliable results. The hierarchical nature of the presentation and the division on sequential species in the plot classification is analogous to the species list approach of the Braun-Blanquet school. There is the same use of species as indicators in a subjective way as in an objective way.

10 Ordination Methods

IN our simple nodum model:

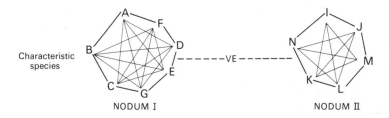

there was a deliberate simplification—implicit in the term nodum, is a 'blurring' of the distinctness of the grouping and:

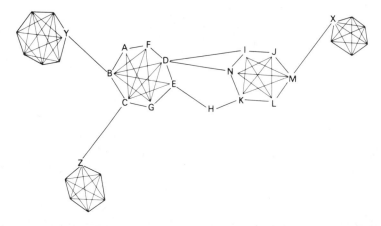

is a much more realistic conceptual model, where A, B, C, etc., are either species *or* plots. There is a spatial configuration implying dimensions as well as a sequence of noda which are not completely isolated but cross-linked. The dimensions can also relate to closeness of the positive associations in terms of whether the clusters are 'tight' or 'loose'. In addition, overall control by environmental factors is implied by the direction of the major axis. This is a simple ordination of species (or plots) and can again be readily obtained by use of χ^2 as a measure of association between species pairs.

The variation of floristic composition of communities dominated by *Juncus effusus* in North Wales was examined by Agnew (1961). He examined 99 stands chosen at random from a potentially large number of areas where *Juncus* occurred. The presence or absence of each species was recorded in each site. Species which occurred less than five times in the total sample were eliminated and 2×2 contingency tables constructed for the remaining 53 species. The correlations obtained (Fig. 10.1) were diagrammatically represented in the form of a two-dimensional spatial arrangement of species (Fig. 10.2).

The reciprocal of the calculated value of χ^2 between each species pair is used to construct the diagram. Thus two species highly positively associated and with a large χ^2-value are positioned close to one another. In practice some approximation is necessary (see below). The arrangement of species reflects positive correlations in the main, the general position of each group of species being related to the negative correlations also present. On pure chance alone some correlations will be significant (1 in 20 at a 5 per cent significance level) and accordingly those species with only a single correlation at a 5 per cent level are not included in the diagram. It is immediately obvious that three groupings of species occur (see Fig. 10.2).

GROUP 1 (left-hand side of the diagram)
> *Holcus lanatus, Lolium perenne, Phleum pratense, Ranunculus repens, Stellaria media* and *Trifolium repens.*

GROUP 2 (lower right-hand side of the diagram)
> *Angelica sylvestris, Carex panicea, Galium uliginosum, Hydrocotyle vulgaris, Lotus uliginosus, Molinia coerulea, Potentilla palustris* and *Succisa pratensis.*

GROUP 3 (upper right-hand side of the diagram)
> *Aulacomnium palustre, Deschampsia flexuosa, Hylocomium splendens, Polytrichum commune, Pseudoscleropodium purum, Rhytidiadelphus squarrosus* and *Sphagnum recurvum.*

It is equally clear that none of the three groupings are isolated and distinct entities, each is linked to the adjacent group by correlations between member species of the respective groups or through an intermediate species, e.g. *Juncus acutiflorus*. Agnew suggests it is possible that such intermediate species may prove on a more extensive survey to be part of additional groupings. The deduction that these correlated groupings of species represent significant 'communities' or plant associations (in

Fig. 10.1. Complete chi-square matrix for 99 *Juncus effusus* stands showing the positive and negative species relationships present. (From Agnew 1961; courtesy of *J. Ecol.*)

Fig. 10.2. The species 'constellation' based on chi-square values and showing positive correlations from 99 *Juncus effusus* stands. (From Agnew 1961; courtesy of *J. Ecol.*)

Acu	*Acrocladium cuspidatum*	Lu	*Lotus uliginosus*
Apa	*Aulacomnium palustre*	M	*Molinia caerulea*
Apt	*Achillea ptarmica*	Ma	*Mentha aquatica*
Asy	*Angelica sylvestris*	Pco	*Polytrichum commune*
Br.	*Brachythecium rutabulum*	Pe	*Potentilla erecta*
Ce	*Carex echinata*	Phl	*Phleum pratense*
Cp	*Carex panicea*	Ppa	*Potentilla palustre*
Cpa	*Cirsium palustre*	Psp	*Pseudoscleropodium purum*
Df	*Deschampsia flexuosa*	Pv	*Prunella vulgaris*
Ep	*Epilobium palustre*	Rf	*Ranunculus flammula*
Fil	*Filipendula ulmaria*	Rr	*Ranunculus repens*
Fr	*Festuca rubra*	Rsq	*Rhytidiadelphus squarrosus*
Gs	*Galium saxatile*	Ru	*Rumex acetosa*
Gu	*Galium uliginosum*	Sm	*Stellaria media*
Hl	*Holcus lanatus*	Smi	*Scutellaria minor*
Hv	*Hydrocotyle vulgaris*	Sp	*Sphagnum palustre*
Hys	*Hylocomium splendens*	Spr	*Succisa pratense*
Ja	*Juncus acutiflorus*	Sr	*Sphagnum recurvum*
Jar	*Juncus articulatus*	Ss	*Sphagnum subsecundum*
Lb	*Lophocolea bidentata*	Tr	*Trifolium repens*
Lp	*Lolium perenne*		

the sense intended by Braun-Blanquet) is considerably strengthened by additional ecological information. Thus the general linear relationship of the three groups follows a pH gradient which is highly significant ($r = -0.509$, $p < 0.001$).

Welch (1960) presents a similar series taken from woodland in Tanganyika (Fig. 10.3). The very obvious groupings on the left of the diagram represents a group of species which form dense thickets, the central group linked to the thicket species through *Albizia, Canthium* and *Acacia pennata* represents a group of pioneer species which invades abandoned cultivated land. Similarly the group of species on the right of the diagram represent a woodland association dominant in some adjacent areas. The three groups in fact reflect the initial and secondary stages of two lines of succession. DeVries (1953) examined a large number of grassland sites in Holland and estimated the percentage frequency of the species in each of the sample sites. Correlation coefficients were calculated and the data presented in the form of a two-dimensional diagram (Fig. 10.4) based on positive correlations similar to the *Juncus effusus* example above. Again a number of groupings are present, all of them inter-related yet forming fairly distinct ecological groupings.

GROUP 1 (top left of the diagram)
Molinia caerula, Sieglingia decumbens, Carex panicea, Potentilla erecta and *Cirsium dissectum* typical of acidic soils of low fertility.

GROUP 2 (bottom left of the diagram)
Glyceria maxima, Caltha palustris, Phalaris arundinacea, Lychnis flos-cuculi, Carex disticha, Ranunculus repens, Glyceria fluitans and *Alopecurus geniculatus*, typical of moist neutral-acidic grassland.

GROUP 3 (top right of the diagram)
Arrhentherum elatius, Trisetum flavenscens, Dactylis glomerata, typical of dry hay meadows.

GROUP 4 (bottom right of the diagram)
Lolium perenne, Cynosurus cristatus, Poa pratense, Agrostis stolonifera, Plantago major and *Trifolium repens*, typical of fertile neutral pastures.

This method of approach has several advantages, the initial subjective choice of site is eliminated to a large extent and a partial-random choice can be made at least. The initial decision of the type of site (a grassland, woodland, etc.) is obviously unavoidable. Secondly the groupings of species are completely objective and present a more realistic picture with the inter-relationships and almost continuous variation of the groupings readily appreciated. At the same time the element of constancy is present and Agnew (1961) suggests *Trifolium repens, Potentilla palustris* and *Galium saxatilis* represent constants for groups 1, 2, and 3 respectively. Also the artificial hierarchical classification is avoided and if a sufficiently

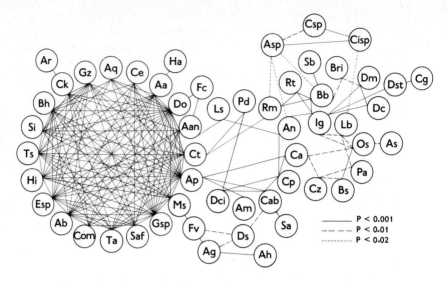

Fig. 10.3. Positive associations between plant species in woodland in Tanganyika.
(From Welch 1960; courtesy of *J. Ecol.*)

Aa	*Allophylus alnifolius*		Dci	*Danbeya cincinnata*
Aan	*Albizia anthelmintica*		Dm	*Dalbergia melanoxylon*
Ab	*Albizia brachycalyx*		Do	*Dalbergia ochracea*
Ag	*Acacia gerrardii*		Ds	*Dombeya shupangae*
Ah	*Albizia harueyi*		Dst	*Dalbergia stuhlmannii*
Am	*Acacia mellifera*		Esp	*Euphorbia spp*
An	*Acacia nigrescens*		Fc	*Fagara chalybea*
Ap	*Acacia pennata*		Fv	*Flueggea virosa*
Aq	*Afzelia quanzensis*		Gsp	*Grewia spp*
Ar	*Asparagus racemosus*		Gz	*Gelonium zanzibarense*
As	*Acacia senegal*		Ha	*Harrisonia abyssinica*
Asp	*Acacia sp*		Hi	*Haplocoelum inopleum*
Bb	*Brachystegia boehmii*		Ig	*Isoberlinia globiflora*
Bh	*Brachylaena hutchinsii*		Lb	*Lonchocarpus bussei*
Bn	*Bridelia niedenzui*		Ls	*Lannea stuhlmannii*
Bs	*Boscia salicifolia*		Ms	*Manilkara sulcata*
Ca	*Combretum apiculatum*		Os	*Ostryoderris stuhlmannii*
Cab	*Cassia abbreviata*		Pa	*Pterocarpus angolensis*
Ce	*Carissa edulis*		Pd	*Phyllanthus discoideus*
Cg	*Combretum ghasalense*		Rm	*Royena macrocalyx*
Cisp	*Cissus sp*		Rt	*Randia taylorii*
Ck	*Capparis kirkii*		Sa	*Steganotaenia araliacea*
Com	*Commiphora spp*		Saf	*Spirostachys africana*
Cp	*Commiphora pilosa*		Sb	*Sclerocarya birrea*
Csp	*Cordia sp*		Si	*Strychnos innocua*
Ct	*Canthium telidosma*		Ta	*Thylachium africanum*
Cz	*Combretum zeyheri*		Ts	*Teclea simplicifolia*
Dc	*Dichrostachys cinerea*			

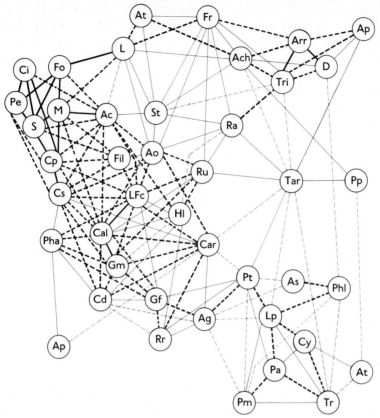

Fig. 10.4. Positive correlations between plant species in grassland, forming a constellation. (From DeVries 1953; courtesy of *Acta Bot. Neer.*)

Ac	*Agrostis canina*	Hl	*Holcus lanatus*
Ach	*Achillea millefolium*	L	*Luzula campestris*
Ag	*Alopecurus geniculatus*	LFc	*Lychnis flos-cuculi*
Ao	*Anthoxanthum odoratum*	Lp	*Lolium perenne*
Ap	*Alopecurus pratensis*	M	*Molinia caerulea*
Arr	*Arrhenatherum elatius*	Pa	*Poa annua*
As	*Agrostis stolonifera*	Pe	*Potentilla erecta*
At	*Agrostis tenius*	Pha	*Phalaris arundinacea*
Cal	*Caltha palustris*	Phl	*Phleum pratense*
Car	*Cardamine pratensis*	Pm	*Plantago major*
Cd	*Carex disticha*	Pp	*Poa pratensis*
Ci	*Cirsium dissectum*	Pt	*Poa trivialis*
Cp	*Carex panicea*	Ra	*Ranunculus acer*
Cs	*Carex stolonifera*	Rr	*Ranunculus repens*
Cy	*Cynosurus cristatus*	Ru	*Rumex acetosa*
D	*Dactylis glomerata*	S	*Sieglingia decumbens*
Fil	*Filipendula ulmaria*	St	*Stellaria graminea*
Fo	*Festuca ovina*	Tar	*Taraxacum officinale*
Fr	*Festuca rubra*	Tr	*Trifolium repens*
Gf	*Glyceria fluitans*	Tri	*Trisetum flavescens*
Gm	*Glyceria maxima*		

widespread survey was made, the *dimensional* inter-relationships of a large number of associations could be readily defined. The major disadvantage of the approach lies in the lengthy computation necessary where the number of sites is large. Since the larger the number of samples the more complete and significant the inter-relationships become this is a serious disadvantage. The lack of over-all sensitivity of the method also detracts from its use as a standard approach.

Although environmental gradients are also implied, the species constellation diagram does not show them clearly. Corresponding plot constellation diagrams are usually too complicated and cumbersome to attempt, even though an environmental gradient could be evidenced as an equivalent abundance gradient of species components in the plot ordination. That species abundance can be equated to environmental parameters is an important concept with far reaching implications (see below).

A more effective way of ordinating plots in relation to an environmental gradient is that developed by Cottam (1949) and Curtis and McIntosh (1950, 1951). This development is usually called the *Continuum approach*. It implies a continuous variation of vegetation, and thus a viewpoint that vegetation is a *continuum* and cannot be classified into discrete entities (associations in the Braun-Blanquet sense). Furthermore rather than sampling the presence or absence of a species as a simple attribute of the vegetation, the method employs abundance measures. The definitive data was obtained from a randomized sample of woodlands in N. America. The choice of a sample site was based on criteria which avoided preconceived ideas and eliminated the conscious selection of sites inherent in the Braun-Blanquet approach. Thus any woodland stand was sampled which was 'natural' (not planted), of adequate size (15 acres being the minimum) and not disturbed as far as could be determined by grazing, excessive cutting or fire. All the species of trees, herbs, shrubs were included and the frequency, density and dominance measured. Cottam (1949) defined frequency as the percentage occurrence of a species in the total sample (p. 16), density as the total number of individuals of a species relative to the total area examined, and 'dominance' is expressed as basal area. From the frequency, density and dominance values a 'relative importance' value was derived by addition of the values, the DFD index (Curtis 1947). Subsequently these measures were modified by Curtis and McIntosh (1950) and used in the form of:

$$\text{Relative density} = \frac{\text{Number of individuals of species}}{\text{Total number of individuals of all species}} \times 100$$

$$\text{Relative frequency} = \frac{\text{Frequency species x}}{\text{Sum of frequency values of all species}} \times 100$$

and similarly relative dominance is used instead of dominance.

These measures are totalled and constitute the *importance value*, which can range from 0 to 300. The subsequent treatment of the data is most suitably described by reference to a specific set of data, given by Brown and Curtis (1952). The data relate to upland Conifer-hardwood forests of northern Wisconsin, the sampling sites being chosen by the criteria listed above with the additional feature of the absence of standing run-off water. For each stand in turn, an importance value is calculated for each species present and by inspection of the importance values each stand is assigned a *leading dominant*, i.e. that species with the highest importance value. Having designated the leading dominant in each stand, the data is then sub-divided so that each group of stands has the same leading dominant. Obviously some of the subordinate species in any one stand will be the leading dominants of other stands and consequently a fresh importance value is then calculated for each species based on the group of stands with their common leading dominant. In the conifer-hardwood data the leading dominants (Table 10.1) are *Tsuga canadensis, Quercus rubra, Betula lutea, Acer rubrum, Betula papyrifera, Pinus strobus, Populus tremuloides, Pinus resinosa, Quercus ellipsoidalis* and *Pinus banksiana* in decreasing order of importance (importance values 25, 22, 21, 7, 6, 1, 1, 0, 0 and 0 respectively). *Acer rubrum* and *Betula lutea* do not achieve dominance in

Table 10.1 Average importance value of trees in stands with given species as leading dominant—104 stands. (From Brown and Curtis 1952; courtesy of *Ecol. Monog.*)

Number of Stand	Leading dominant	Acer saccharum	Tsuga canadensis	Betula lutea	Acer rubrum	Quercus rubra	Betula papyrifera	Pinus strobus	Pinus resinosa	Populus tremuloides	Quercus ellipsoidalis	Pinus banksiana
23	Acer saccharum	145	25	21	7	22	6	1	–	1	–	–
23	Tsuga canadensis	40	152	47	11	3	5	4	3	–	–	–
6	Quercus rubra	27	1	3	29	138	23	10	8	5	3	–
6	Betula papyrifera	48	8	7	27	16	108	19	1	29	1	–
19	Pinus strobus	12	6	2	24	12	12	150	39	9	5	–
9	Pinus resinosa	3	–	1	12	15	14	56	156	24	4	2
4	Populus tremuloides	11	–	–	10	29	34	14	19	140	–	–
4	Quercus ellipsoidalis	–	–	–	5	7	1	11	9	9	103	56
10	Pinus banksiana	–	–	–	3	3	3	13	12	14	36	213

any stand and accordingly are not included by Brown and Curtis in the provisional arrangement of 'leading dominants'. These two species considered apart, the remaining mean importance value in those stands dominated by *Acer saccharum* give a general basis for the arrangement of the leading dominants in a phytosociological order. Thus *Pinus banksiana* has an importance value of o and is generally ecologically accepted as a typical pioneer species in Wisconsin, the remaining species falling between this and the final 'climax' association dominated by *Acer saccharum*.

The next highest importance value to *Acer saccharum* is *Tsuga canadensis* and accordingly this is placed in an adjacent position to *Acer saccharum*. Confirmation of the validity of this is obtained from the relative size of the mean importance value of the species in the *Tsuga* dominated group. Thus (discounting *Betula lutea*) *Acer saccharum* has the highest importance value and should thus be close to the group which is in fact dominated by *Acer saccharum*. On this basis and with due consideration to the columns of numbers both horizontal and vertical, approaching at least a smooth curve, the remaining leading dominants are arranged. On the basis of the order of the leading dominants the remaining tree species are included, and an arbitrary climax adaptation number ranging from 1 to 10 is assigned to each species (Table 10.2). *Pinus banksiana* is given an adaptation number of 1 and similarly *Acer saccharum* 10 representing the two ends of the continuum. The remaining species receive adaptation numbers of intermediate value so that species which frequently occur together and have similar environmental requirements, have the same or closely similar adaptation numbers. Finally the importance value of each species in each sample is multiplied by the adaptation number, the total sum of these values for each stand being termed the *continuum index* of the stand. This index is regarded as a measure of the total environment of a stand of trees expressed in terms of species composition and their relative abundance. Brown and Curtis (1952) give a detailed table of data (Table 10.3) representative of the information extracted by the method from a mass of otherwise extremely variable data. The results are best appreciated as a series of graphs. A considerable scatter of points is usually present and the data is 'smoothed' by a standard procedure given by the formula

$$B = \frac{n_1 a + 2n_2 b + n_3 c}{n_1 + 2n_2 + n_3}$$

where B is the smoothed value of average b and n_1, n_2 and n_3 are the numbers of stands from which the averages a, b and c have been derived. The relationship between importance value and continuum index is a Gaussian type of curve or part of such a curve (Figs. 10.5 and 10.6), and clearly there is no grouping of curves in zones of the continuum index as should be the case if there were discrete and recognizable associations. On the

Table 10.2   Climax adaptation numbers of tree species found in stands studied.
(From Brown and Curtis 1952; courtesy of *Ecol. Monog.*)

Tree species	Climax adaptation number
Pinus banksiana	1
Quercus ellipsoidalis	2
Quercus macrocarpa	2
Populus balsamifera	2
Populus tremuloides	2
Populus grandidentata	2
Pinus resinosa	3
Pinus pensylvanica	3
Quercus alba	4
Prunus serotina	4
Prunus virginiana	4
Pinus strobus	5
Betula papyrifera	5
Juglans cinerea	5
Acer rubrum	6
Acer spicatum	6
Fraxinus nigra	6
Picea glauca	6
Quercus rubra	6
Abies balsamea	7
Thuja occidentalis	7
Carpinus caroliniana	7
Tsuga canadensis	8
Betula lutea	8
Carya cordiformis	8
Fraxinus americana	8
Tilia americana	8
Ulmus americana	8
Ostrya virginiana	9
Fagus grandifolia	10
Acer saccharum	10

* Climax adaptation number of these species is tentative only, because of their low presence in the stands studied.

contrary there is always a continuous overlap of several adjacent leading dominants. Similarly the species with intermediate and with lower importance values show the same overlap over a considerable range of the

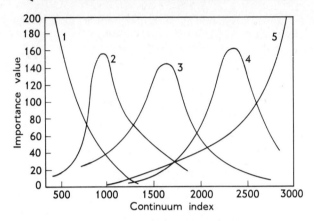

Fig. 10.5. Importance value curves for the five leading tree species (1) *Pinus banksiana*, (2) *P. resinosa*, (3) *P. strobus*, (4) *Tsuga canadensis*, (5) *Acer saccharum*. (From Brown and Curtis 1952; courtesy of *Ecol. Monog.*)

continuum index. It is of considerable interest that measures of environmental factors taken simultaneously with sampling of each stand show marked trends. The specific gravity of screened soil samples, the moisture holding capacity and the available calcium, in the A_1 horizon all follow a very definite trend in relation to the continuum index (Figs. 10.7, 10.8 and 10.9). In addition to the inter-relationship of the species, some interactions between the vegetational continuum and the environment are thus suggested.

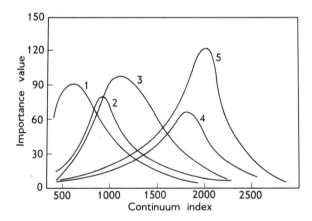

Fig. 10.6. Importance value curves for five intermediate tree species. (1) *Quercus ellipsoidalis*, (2) *Populus tremuloides*, (3) *Quercus alba*, (4) *Betula papyrifer*, (5) *Quercus rubra*. (From Brown and Curtis 1952; courtesy of *Ecol. Monog.*)

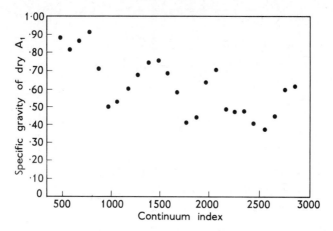

Fig. 10.7. Specific gravity of screened samples of the A_1 soil horizons of stands arranged in 100 unit classes. (From Brown and Curtis 1952; courtesy of *Ecol. Monographs.*)

The assignment of the climax adaptation number is necessarily partly subjective and can be criticized on these grounds. However, considerable use is made of the relative importance values of each species within each leading dominant group. Data obtained from limestone grassland in Yorkshire is given below (Tables 10.4, 10.5 and 10.6) to illustrate the point. The data were obtained from a number of sites, frequency and cover repetition (p. 16) being recorded for each species and used jointly as a

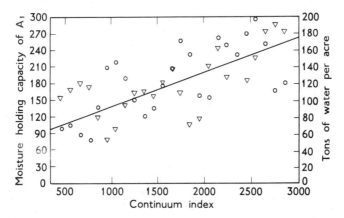

Fig. 10.8. Moisture holding capacity of the A_1 horizon in percentage, and total water held in the entire A_1 horizon at field capacity in tons per acre. Points are average values in 100 unit classes. (From Brown and Curtis 1952; courtesy of *Ecol. Monographs.*)

Table 10.3   Original data for importance values of seventeen species and for three soil characters. (From Brown and Curtis 1952; courtesy of *Ecol. Monog.*)

Stand Number	Continuum Index	*Pinus banksiana*	*Quercus ellipsoidalis*	*Populus tremuloides*	*Populus grandidentata*	*Pinus resinosa*	*Quercus alba*	*Pinus strobus*	*Betula papyrifera*	*Acer rubrum*	*Quercus rubra*	*Abies balsamea*	*Betula lutea*	*Tsuga canadensis*	*Ulmus americana*	*Tilia americana*	*Ostrya virginiana*	*Acer saccharum*	Calcium in A_1 layer (100's lb. per acre)	pH of A_1 layer	WRC of A_1 layer
084	356	272	4	9	6	12	5			4									4.3	5.1	77
066	375	237	46	7			16	5	15	14									4.3	4.9	36
068	467	164	117	4															2.9	4.6	54
098	560	157	95	14		46													2.1	4.6	164
055	594	178		30	7			39											1.4	4.5	102
049	664	122	100				62	6		11									2.9	4.0	118
082	721	181		4	8	34		59	4	4		4							1.1	4.6	99
067	749	100	62	9		7	101	14	3		5								2.9	4.7	54
113	848	10		68		204													6.4	4.3	148
064	868			194	24	7		34	16	14	4								—	—	—
039	934		132		27	268	107	16	4	3	22								2.1	4.7	60
115	954		6	54	4	5			12	35	3								7.2	5.4	110
150	984			63	162	159		74	30										—	—	—
053	991			62	5	116		121											—	4.3	400
140	1082								4	3									3.6	5.1	281
044	1107		8	4	20	153	11	82	3		18	3		3					—	—	—
141	1168	7	4	31	33	65		145		3	7								3.6	5.2	400
132	1209				5	138		157											0.7	4.5	269
059	1290		5	4		96		186	7										—	—	—
070	1363		32	16			109	46		84									4.3	5.7	43

ID	N			
050	1403	7·1	4·9	49
089	1436	0·7	5·2	126
107	1508	6·4	5·2	106
106	1518	8·6	5·5	180
090	1573	3·2	5·5	65
105	1582	8·6	5·0	225
043	1640	5·4	4·9	89
147	1750	5·4	5·2	272
153	1807	2·1	5·3	288
086	1888	0·4	4·0	192
102	1950	2·7	4·9	91
095	1966	1·1	5·3	67
101	2060	12·5	5·5	80
119	2173	35·7	6·1	373
139	2215	3·2	4·6	270
138	2234	2·1	3·7	409
104	2287	16·0	4·5	289
116	2314	7·2	5·2	209
111	2366	8·6	5·0	223
063	2379	—	—	—
136	2404	5·4	4·6	400
085	2430	14·3	5·5	258
062	2464	7·1	5·6	210
134	2502	4·3	4·4	335
118	2520	12·9	5·0	497
125	2553	3·6	5·4	398
034	2591	45·7	5·6	365
121	2617	28·6	5·4	131
103	2636	42·8	5·5	305
135	2711	42·9	6·9	119
080	2747	6·4	4·8	158
094	2780	23·6	5·3	129
076	2817	51·0	7·6	137
133	2852	10·7	5·7	526
114	2938	10·0	5·0	157

Fig. 10.9. Available calcium in the A_1 horizon in pounds per acre. Points are average values in 100 unit classes. (From Brown and Curtis 1952; courtesy of *Ecol. Monographs.*)

measure of the importance value. The limestone area is very variable, the soils ranging from pockets of acidic boulder clay to dark soils immediately overlying limestone rock with a pH above 7·0. The grass cover reflects this wide range of pH, varying from *Nardus stricta* dominated areas to almost pure stands of *Sesleria caerulea*. The importance value calculated from the relative frequency and relative cover of each species, shows *Festuca rubra*, *Agrostis tenuis*, *Nardus stricta* and *Sesleria caerulea* to be the leading dominants in the area. The data are accordingly split into these four groups and importance values calculated for each species (Table 10.4).

The leading dominants ecologically form a series with *Nardus* and *Sesleria* at the two extremes and *Agrostis/Festuca* falling in an intermediate position. The importance values in each group confirm this relationship *Agrostis* and *Festuca* having a similar relationship at the acidic end of the scale, but *Festuca* being clearly more closely related to *Sesleria* at the basic

Table 10.4—Importance values of the four most abundant species in relation to the leading dominants in an area of grassland in Yorkshire

		Nardus	*Agrostis*	*Festuca*	*Sesleria*
	Nardus	58·6	44·0	49·5	0
Leading dominant	*Agrostis*	36·0	58·1	50·2	4·5
	Festuca	15·5	37·5	53·9	12·7
	Sesleria	0	13·6	45·1	61·6

Table 10.5   The importance values of each species present in an area of grass-land overlying limestone in Yorkshire in relation to four leading dominants

Species	Festuca group	Agrostis group	Sesleria group	Nardus group
Festuca rubra	53·9	50·2	45·1	49·5
Agrostis tenuis	37·5	58·1	13·6	44·0
Sesleria caerulea	12·7	4·5	61·6	0
Nardus stricta	15·5	36·0	0	58·6
Koeleria gracilis	9·3	0	9·7	0
Thymus drucei	3·8	3·6	11·7	0
Linum catharticum	2·5	0	11·2	0
Viola riviniana/lutea	5·8	3·6	11·4	0
Carex flacca/panicea	11·9	7·9	18·7	0
Deschampsia caespitosa	3·9	7·9	0	0
Trifolium repens	6·2	0	5·5	0
Potentilla erecta	8·4	0	0	0
Anthoxanthum odoratum	5·1	0	1·0	0
Galium hercynicum	10·6	22·8	0	22·0

end. The remaining species are included in the comparison and import-ance values calculated for each species in each group (Table 10.5). From Table 10.5 it is now possible to arrange all the species in their relative order and assign to each of them an adaptation number. *Nardus* is arbitrarily de-signated 1 and is most closely related to *Agrostis*, *Galium hercynicum* is similarly closely related to both *Agrostis* and *Nardus* and in Table 10.5 shows a slightly closer relation to *Agrostis* than to *Nardus*. Accordingly *Galium* is allocated an adaptation number of 3 and *Agrostis* 4 (Table 10.6). The species of maximum importance left in the *Agrostis* group are *Carex* and *Deschampsia* and reference to the *Festuca* group shows *Deschampsia* to be much more closely related to *Agrostis* and accordingly is allocated 5 as

Table 10.6

Species	Adaptation number
Nardus stricta	1
Galium hercynicum	3
Agrostis tenuis	4
Deschampsia caespitosa	5
Anthoxanthum odoratum	6
Trifolium repens	6
Potentilla erecta	7
Festuca rubra	7
Koeleria gracilis	8
Viola rivinina/lutea	8
Linum catharticum	8
Thymus drucei	8
Carex flacca/panicea	9
Sesleria caerulea	10

its adaptation number. At the other end of the series *Sesleria* is arbitrarily designated as 10, and clearly *Carex* is closely related to it and allocated an adaptation number of 9. Similarly *Thymus, Linum, Viola* and *Koeleria* are more or less equally related to *Sesleria* and are thus all given an adaptation number of 8. *Festuca* is allocated an adaptation value of 7 and the remaining three species referred to a value of 6. These three species, *Anthoxanthum, Trifolium* and *Potentilla,* all have low importance values and are unfortunately of relatively infrequent occurrence in the samples available. It is possible that a larger sample might indicate a reversal of the order of *Festuca* and these three species, but it is unlikely that the present arrangement will unduly influence the continuum indices of the samples. This final arrangement of species is given in Table 10.6.

Thus the allocation of an adaptation number is based largely on the relative orders of the importance values of each of the species and the element of subjective assessment necessary is slight and restricted to the species with a low percentage occurrence. This can be circumvented by taking a more adequate sample but the labour of such an increase in sample size could be excessive in some instances. The general relationship of the four most abundant species again shows a considerable overlap of their range (Fig. 10.10) relative to the continuum index and is broadly similar to the results given by Brown and Curtis (1952). There is no discrete zone solely dominated by one species, or a group of species, but instead there is a continuous variation of composition along the continuum index.

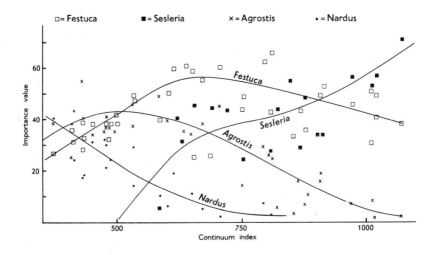

Fig. 10.10. The importance value/continuum index curves for the four leading dominants of sample plots in limestone grassland.

In essence the continuum approach is using a single axis which is proportional in length to the abundance of the commonest (dominant) species in the sample. Each plot sample is then positioned on that axis. The so-called climax adaptation numbers are simply *weighting factors* to 'magnify' the effect of inter-plot abundances and thus distances. All the species are used to assess inter-plot distance and where there is a *single over-riding environmental parameter* controlling the vegetation, only then will the continuum approach produce a sensible result. However, when two or more separate parameters are controlling the plant associations sampled, a single axis will not form an effective framework on which to ordinate the samples. Two or more axes will be required.

Using a simple vector representation, two environmental parameters A and B, that operate on vegetation independently can be represented as:

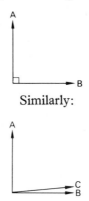

Similarly:

represents three environmental parameters A, B and C where B and C are closely correlated together but both unrelated to A. For a series of plant associations 'A' could be pH, 'B' aspect and 'C' temperature, and correlated with each axis would be a range of species abundances which in turn were controlled by the parameters pH and temperature. For example:

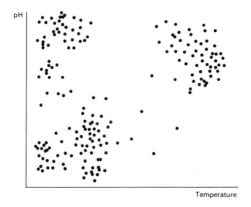

In this simple ordination model we have a number of plots on warm (south facing) slopes with high pH, a group with cooler temperatures and with high pH, and a large group of low pH, low temperature plots. A simple ordination using the continuum approach would give a very distorted image of the real situation. Since many environmental factors operate simultaneously on vegetation and only in a gross climatic sense are there usually over-riding parameters, the continuum as a method will have a very restrictive usage; effectively only in geographical surveys where a single dominant factor is operating.

In the simple ordination model used above, instead of the axes being pH and temperature respectively they could be two of the commoner *species* whose distribution is controlled by pH and temperature. In general, ordination axes are usually related to the abundance of 'indicator' species, where hopefully, information about the ecology of each species is available and can be used to identify the environmental variables correlated with each axis. Often the correlation of an axis with an environmental variable is intuitive and necessitates a considerable amount of experimental and environmental work subsequently.

Our model is still simple. The axes are arbitrarily at right angles to each other by convention, there are only two of them, they lie on a single plane surface, and they are straight lines. This is a closely similar model to that used previously (p. 191 above). In addition, as was pointed out above, in the construction of a constellation diagram from the scaled reciprocals of the χ^2 values between species pairs, some approximation is necessary. This arises during the construction of the diagram in the following way.

If A, B and C are equally correlated together then this relationship can be simply represented as:

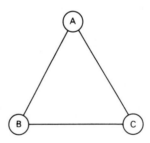

When a fourth species D is introduced, also equally correlated with A, B and C, there is no simple way this can be represented in the diagram. A three-dimensional model is required and the diagram can only (at best) represent a three-dimensional arrangement of species, condensed into two dimensions:

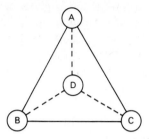

When further species are added to the diagram, numerous other distortions are necessary resulting in the necessary approximations of inter-species distances. This is a general situation. To accurately represent inter-correlation distances between either species or plots, necessitates many dimensions. Three species or plots can be represented in two dimensions, four requires three dimensions and so on. The difficulty of representation in more than three dimensions at once (other than in a mathematical sense) necessitates some simplification and usually only three axes are considered at any one time.

Thus our ordination model can now become:

An example of such an ordination is given below (Fig. 10.11) of 130 plots sampled from lichen heath in Northern Ontario. The three species chosen as axes were three of the most abundant in the area sampled. *Cetraria nivalis* is representative of slightly moist conditions, *Cladonia rangiferina* characteristic of the general associations on the intermediate and dry slopes, and *Dryas integrifolia* dominant on the exposed and dry summits of mounds and low ridges. As a dimensional ordination approach its chief merit is its extreme simplicity. It is a gross approximation since only *three* of the N-species are being used. Having spent considerable time and effort sampling the total species complement, throwing most of it away subsequently is totally unacceptable. Even more serious, information relating to the structure of the ordination, carried by the other common species, is also being thrown away. The extent of this information loss can be seen by comparing Figs. 10.11 and 10.12 which is an analysis of the same lichen heath data using a component analysis (see below).

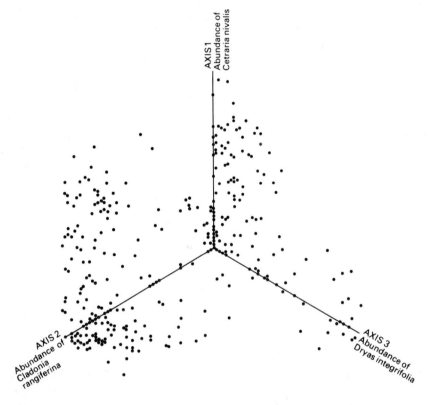

Fig. 10.11. Ordination of plots using three of the commonest species in the data as abundance axes. The data taken from lichen heath in Northern Ontario.

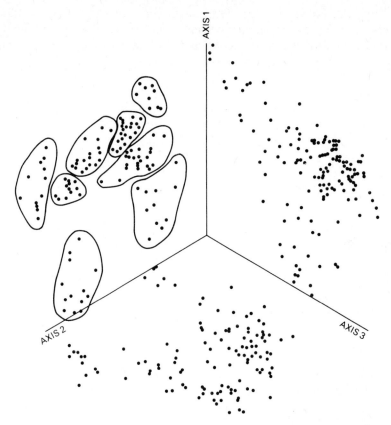

Fig. 10.12. Plot ordination of the same lichen heath data from Northern Ontario, but using a principal component analysis. The ringed points represent the superimposed Braun–Blanquet associations previously extracted.

We arrive then, at our final model situation for a plot ordination:

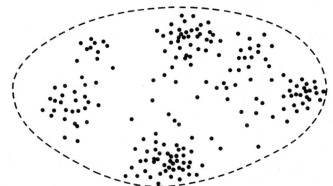

We know the sample plots lie in some type of spatial cluster of ellipsoidal form. We do not want to use simple species axes, because this is too wasteful. But certainly axes *are* required to which the data can be related. Three possible configurations could be:

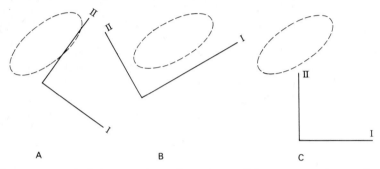

A   B   C

There are an infinite number of such possible axes to choose from. Intuitively 'B' would seem to be better than 'A' or 'C'. but the objective reasons for this decision involve some complex mathematical concepts and techniques.

One simple approach is that given by Bray and Curtis (1957), using a series of coefficients of similarity to set up a primary and subsequent continuum axes. The sampling sites were chosen at random as before and quantitative measurements taken of the tree, shrub and herb population in each stand (density and basal area for the trees, frequency for the herbs and shrubs). Fifty-nine stands were sampled and each of these stands was compared with the other 58 and a coefficient of similarity calculated. The coefficient used was that proposed by Gleason (1920), $C = 2w/(a+b)$, where a and b are the quantities of all the plants found in the two stands to be compared and w is the sum of the lesser value for those species common to the two stands. Thus if the two stands were absolutely identical a and b and w would be equal and C then equals 1. Conversely with complete dis-similarity $C = 0$. From the 1711 values thus obtained the two most dis-similar stands are chosen as the end points of an axis and the remainder of the stands located on the axis by using the inverse value of the other coefficients as ordinates. From this arrangement several stands were found to be more or less equidistant from the two ends of the first axis but were still very dis-similar. These were examined and the most dis-similar pair again used to establish a second axis. The stands were re-positioned by the two axes, again using the inverse coefficients as co-ordinates. A third axis was finally established on the most dis-similar pair of stands which occurred equidistant from the ends of the axes but which were still very different in composition. From the resultant three axes, each stand is thus located by three co-ordinates to form a three-dimensional spacing. This final positioning in three dimensions produced a highly significant correlation between the spacing of the stands and the original coefficient calculated from the data. From the spatial arrangement of the stands the arrangement of the species can be readily determined although there is some difficulty in representing the three-dimensional spacing on paper. Bray and Curtis give data of the dominant tree species (Figs. 10.13 and 10.14). Each species shows a definite spatial 'centre of importance' (related to the number and size of the balls). Away from the centre the concentration and importance of a species diminishes, but not equally in different directions. The three axes are interpreted as: 1. Reflecting recovery from major past disturbance and shows a marked correlation with features of an increasingly mesic environment and with several soil factors—organic matter, pH, Ca, P, etc. 2. Related to drainage and soil aeration and shows correlation with soil water retaining capacity and ammonium ion concentration. 3. Related to recent disturbance and the influence of gap phase replacement.

The ordination procedure of Bray and Curtis (1957) is fairly crude and very unwieldy with a large sample. It did, however, stimulate a considerable interest in ordination methods in general, and specifically the use of more exact and effective methods of axis construction. In 1954, Goodall had applied the technique of factor analysis to ordinate Australian mallee data. The extremely heavy computation necessary certainly limited the impact of the paper. Dagnelie (1960) extended its application further and with the increasing availability of computers a number of related techniques have now been employed. The methods can be considered as two major categories from a user's point of view. Factor analysis which attempts to look for groups of correlated environmental parameters controlling different plant association, and component analysis which seeks to display, in as economical a way as possible, the continuous variation of population structure. The mathematically inclined reader is referred to Williams and Dale (1965) and Grieg-Smith (1964) for the general implications in classification and ordination approaches, and Kendall (1957) for computational details. The discussion here will be restricted to component analysis which as an *ordination technique* is extremely efficient.

More simply from our final ordination model:

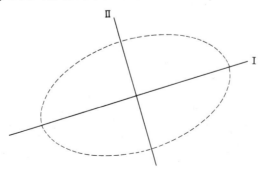

component analysis would construct axis I through the maximum spatial variation present, axis II through the next highest spatial variation and so on, each axis being at right angles to the previous axis. Component analysis achieves, with maximum efficiency and in an objective way, what was

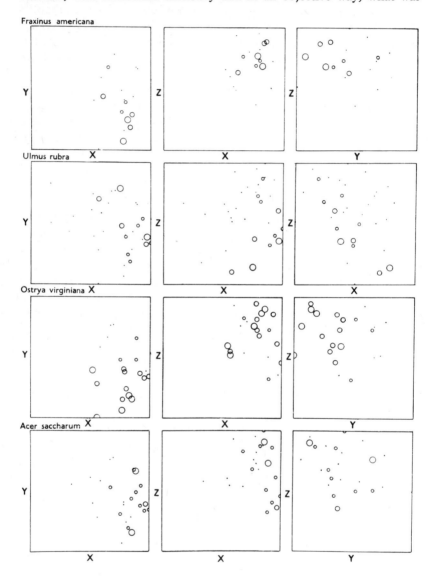

Fig. 10.13 and **10.14.** (*opposite*) The dominance behaviour of the eight most important tree species at three levels of the ordination. (From Bray and Curtis 1957; courtesy of *Ecol. Monographs.*)

attempted intuitively (p. 214 above). Since variation in a population is related to one or several environmental parameters and in turn these are reflected in the abundance of indicator species, the resultant ordination can efficiently summarize a large body of information. If noda are present, they equally will be clearly evident in the resultant ordination. Component analysis can operate on a correlation matrix and produce either an ordination of species or plots. Again there has been tremendous confusion over terminology and in agreement with Ivimey-Cook *et al.* (1969) a

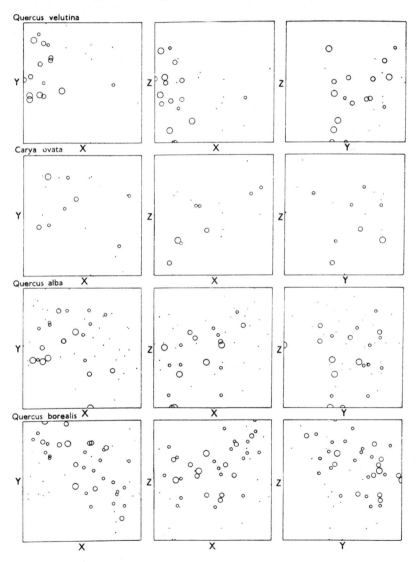

descriptive terminology is recommended (see p. 189 above). Computer facilities are essential for a component analysis and several computer packages are now available. The output is in the form of a series of eigen values (latent roots) which are proportional to the variation accounted for by each axis, with eigen vectors (latent vectors) which are the loadings or, after standardization in each set so that the sum of their squares is equal to the respective eigen value, simply the spatial co-ordinates of each plot (or species in a species ordination). Usually only the first three axes are examined, the fourth and subsequent axes only accounting for small and decreasing amounts of residual percentage variation. Most packages will operate on either correlation, covariance or a defined 'similarity' matrix, of which several have been proposed. The choice as to which is most effective has very largely been decided by the outcome of the analysis— from experience a simple correlation coefficient or Orloci's coefficient

$$\sum_{i=1}^{N} (x_{ij} - \bar{x}_i)(x_{ih} - \bar{x}_i)$$

where N = number of species x_i, x_{ij} and x_{ih} are the abundance of species i in samples j and h, and $\bar{x}_i$ is the mean abundance value for species i. Orloci (1966), Austin and Orloci (1966), appear to be equally satisfactory in practice, over a range of data.

The presentation of the data output can take the form of a standard ordination on each of the three pairs of axes, or more effectively as Grieg-Smith *et al.* (1967) have done, as projections of the spatial cluster of plots on each plane surface in turn (Fig. 10.15) (see also p. 241 below).

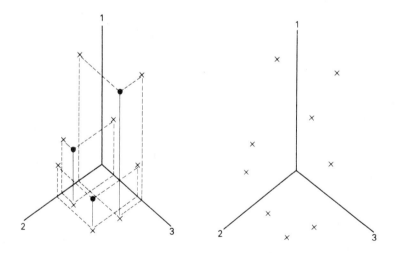

Fig. 10.15. Method of projecting the spatial plot locations on to the three plane surfaces composed of Axis 1-2, Axis 1-3, and Axis 2-3.

The identification of the trends in the ordination is achieved by over-laying species abundance values or environmental measures on the plot ordination. Austin and Orloci (1966) ordinated dune slack vegetation. This was from Newborough Warren, Anglesey sampled using percentage frequency as a measure of abundance. The ordination is given below (Fig. 10.16) together with the corresponding plot frequencies of *Ammophila arenaria*, *Potentilla anserina*, and *Carex serotina*. These three species are characteristic of dunes and dry slacks, moist slacks, and wet alkaline slacks respectively. Accordingly axis I is related to a vegetational gradient from transitional dunes or dry slacks, to slack communities, whilst axis II follows a vegetational gradient from wet alkaline slacks to the more leached meadow type of dune slack. Similarly Grieg-Smith *et al.* (1967)

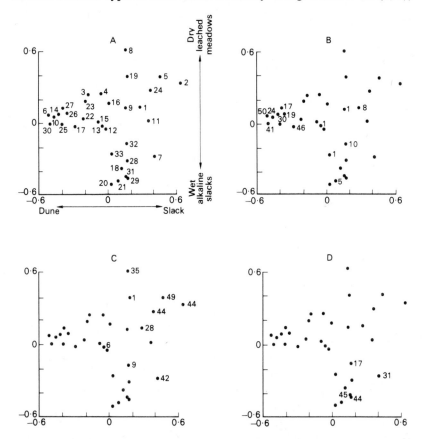

Fig. 10.16. The ordination of dune slack data on the first two component axes (A) with the abundance (per cent frequency) of *Ammophila arenaria* (B), *Potentilla anserina* (C) and *Carex serotina* (D) in each plot. (From Austin and Orloci 1966; courtesy of *J. Ecol.*)

ordinating tropical forest from density data sampled from Kolombangara Island in the Solomon archipelago in the S.W. Pacific showed a marked distribution of altitude and slope classes with the plot ordination (Fig. 10.17).

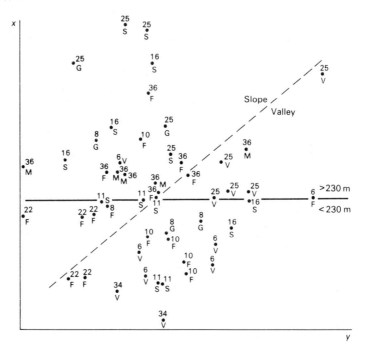

Fig. 10.17. Correlation of altitude and topography with the plot ordination on the first two component axes from tropical forest data in the Solomon Islands. S, Steep slope; G, gentle slope; F, flat/plateau; V, valley; M, varied topography. Figures indicate actual plot altitude in 10-m classes. (From Greig-Smith *et al.* 1967; courtesy of *J. Ecol.*)

Kershaw (1968) ordinating Domin values sampled from savanna vegetation in Northern Nigeria showed the first three component axes were identified with low/high water availability, presence/absence of fossil ironstone cuirass material, and a northern/southern trend respectively. Traditional Braun-Blanquet associations extracted previously showed a close level of agreement with the ordination (Fig. 10.18).

The successful outcome of a component analysis is very dependent on the initial degree of heterogeneity in the data. If the data is markedly discontinuous, for example, with one or two associations which are very clear cut from the remaining central mass of data, these groupings will certainly be distinct in the ordination but the main bulk of the data will also remain as an unstructured 'central mass'. Use of component analysis should be thus restricted to relatively homogeneous data. If the analysis is unsatisfactory then the variation should be reduced. This can be achieved in several ways: extreme associations can be simply deleted from

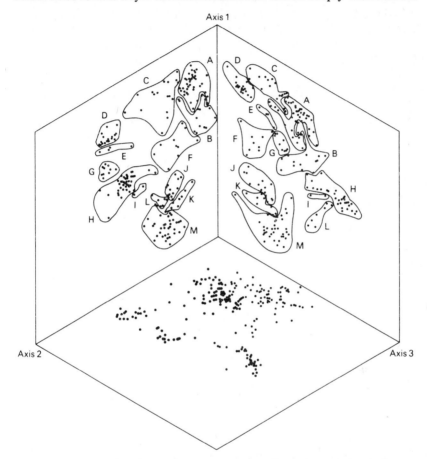

Fig. 10.18. Braun–Blanquet associations superimposed on the principal component plot ordination of savanna data from Northern Nigeria. (A), *Monotes* association (southern); (B), *Monotes* association (northern); (C), *Monotes–Parinari* association; (D), *Parinari–Gardenia–Detarium* association; (E), *Daniellia–Gardenia–Detarium* association; (F), *Monotes–Isoberlinia* association; (G), *Detarium–Gardenia* ironstone association; (H), *Combretum glutinosum–Dichrostachys–Entada* ironstone pavement association, *Anogeissus–Feretia* river bank association, *Bombax–Diospyros–Ficus* inselberg association, and the *Dichrostachys* erosion complex association; (I), flood plain association (J), *Isoberlinia–Parinari* association; (L), *Isoberlinia tomentosa–I. doka* association; (M), *Isoberlinia–Annona–Terminalia–Ximenia* association. (From Kershaw 1968; courtesy of *J. Ecol.*)

the data and the residual bulk of the data re-analysed. Thus Austin and Grieg-Smith (1968) used an association analysis on their Kolombangara Island data to divide it into more homogeneous units prior to component analysis. Kershaw (1968) used square-root transformation of the data and achieved an improvement in the ordination. Thus it is decidedly more effective to analyse presence–absence data, or Domin values, than density or percentage cover data where the variance is inevitably much greater. Implied in this is also the additional advantage of the simplicity of sampling in the field.

Reduction of the total variation in the data is particularly effective in improving a species ordination and especially the removal of the less common species from the analysis.

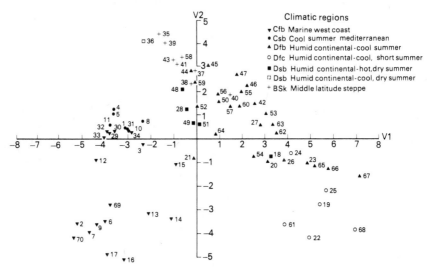

Fig. 10.19. Ordination of meteorological data from 70 stations in British Columbia on the first two principal component axes. (From Newnham 1968; courtesy of *Forest Sci.*)

Thus most of the information relating to the structure of the ordination is carried by the most abundant species. This fact is of considerable importance and immediately simplifies the original collection of the data enormously—the rare species can simply be ignored. Austin and Grieg-Smith (1968) simultaneously reached the same conclusion and showed that an efficient ordination could be achieved with 25 per cent of the total flora. Using standardization procedures, species ordination was markedly improved, especially where the abundance gradient was removed, which usually is extracted by the first axis of a species ordination. Thus component analysis extracts the maximum variance with each axis. In data which is markedly heterogeneous the maximum variance lies in the

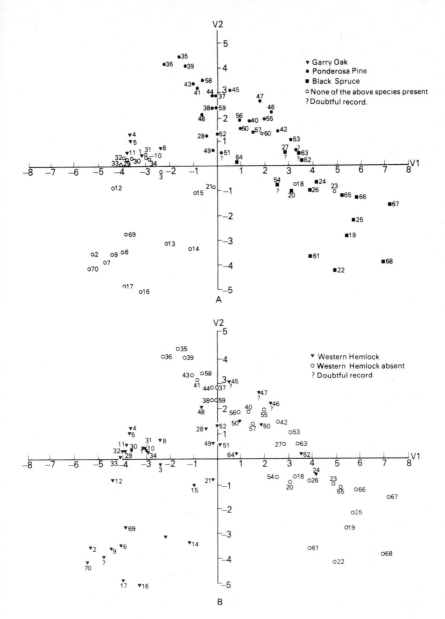

Fig. 10.20. (A) the correlation of Garry oak, Ponderosa pine, and Black spruce distribution with the meteorological ordination; (B) the correlation of Western hemlock and (C) the correlation of Sitka spruce; Engelmann spruce or White spruce distribution with the principal component ordination. (From Newnham 1968; courtesy of *Forest Sci.*)

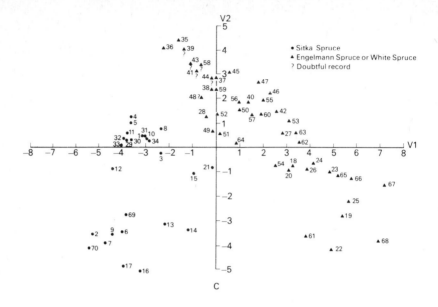

relative species abundances which will range from high levels of abundance to isolated occurrences. The first axis accordingly has the common species loaded at one end and rare species at the other and is of little information value.

Newnham (1968) applied component analysis to climatic data consisting of 19 variables from 70 weather stations in British Columbia. Two components extracted from a correlation matrix accounted for 87 per cent of the total variation. They were correlated with a general index of winter/fall climate/length of growing season, and spring/summer temperatures/precipitation (Fig. 10.19).

Overlaying the known distribution of Sitka spruce, Engelmann spruce or White spruce, Black spruce, Ponderosa pine, Garry oak, etc., reveals a very marked correlation with the ordination of climatic factors (Fig. 10.20). This is an unusual and interesting use of component analysis—in a normal ecological context rarely are there more than 3 or 4 environmental parameters available to correlate with a plot ordination.

The presentation of component analysis output has followed the standard approaches which have been adopted by various workers (Figs. 10.16–10.18). Both these procedures, in fact, obscure the multi-dimensional structure derived from the component analysis and certainly a three-dimensional physical model is considerably more informative. However, the construction of such models is extremely time consuming and only

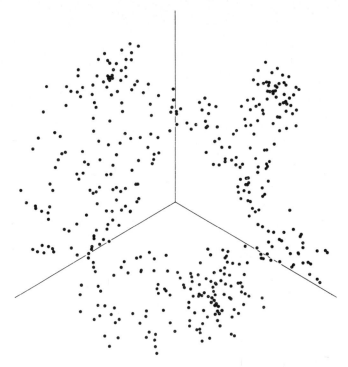

Fig. 10.21. (A) Ordination of lichen heath data from N.W. Ontario as a projection on to the plane surfaces bounded by the first three component axes; (B), (C) and (D) perspective drawings of the spatial configuration of the same data viewed from three different directions.

possible with relatively small samples. The approach of Kershaw and Shepherd (1972), utilizing the off-line graphics facilities of a computer to output perspective drawings of the three-dimensional spatial configuration as viewed from any desired observer's location, is an improvement (see also p. 241 below). Considerable structural detail emerges which is completely obscured by projection of the spatial clusters on to the plane surfaces 1–2, 1–3, and 2–3 (Fig. 10.21). Thus where such facilities can be accessed it is advisable to follow this approach.

SUMMARY AND CONCLUSIONS

Although constellation diagrams offer a simple method of approach to an objective ordination, with large numbers of species or plots, they become very impractical. The technique is insensitive and only utilizes one parameter of the population structure—presence and absence. The continuum approach, although utilizing species abundance values, is only

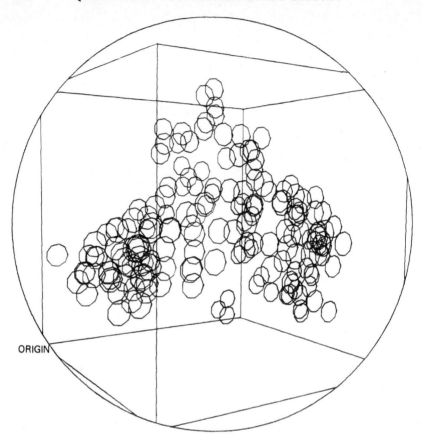

Fig. 10.21. (B) The origin of the ordination lies at the left centre (ORIGIN) of the model.

valid for single, major environmental gradients. Since there are usually many environmental parameters involved, it is not effective in many situations. Use of the common species themselves as both environmental 'indicators' and axes, although introducing a multi-dimensional concept in ordination, ignores the information carried by the other common but unused species in the population. Component analysis is a very flexible technique which objectively and with maximized efficiency utilizes all the correlations present within a population of plots or species, to construct ordination axes. These axes relate sequentially to variation extracted. Plot ordinations are little affected by either data standardization or the spectrum of species abundance. Species ordinations conversely are

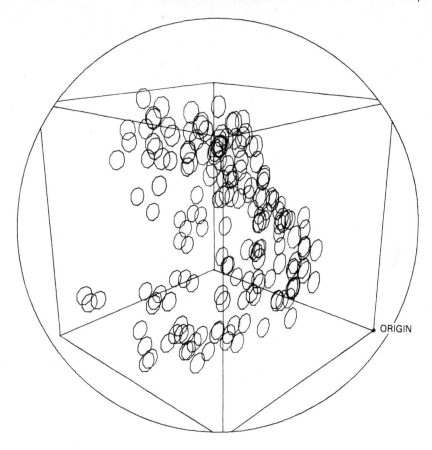

ORIGIN

Fig. 10.21. (C) Perspective view of the same model viewed from 'behind'.

very sensitive to these aspects of data structure. Species abundances overlayed on the plot ordination indicate specific trends in the ordination, and equally environmental measures can be used in a similar way.

It is important to appreciate that component analysis has strict limitations and is *one model* to which data can be fitted. Just as the simple models used above, to develop an appreciation of the current multidimensional approaches in use, contained a number of (loosely) defined properties, component analysis also has a series of rigid properties. One of these properties is rarely found in an ecological situation—orthogonal axes. Axes that are orthogonal simply have no correlation existing between them. Thus a north–south effect in Nigerian savanna *is* in effect correlated

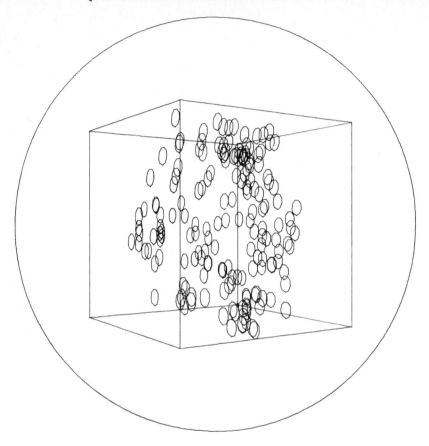

Fig. 10.21. (D) Perspective view of the model after a further rotation, the origin now lying at the right rear of the ordination.

strongly with water availability. The rainfall gradient runs almost exactly N–S. The effect of using orthogonal axes in this type of situation is sometimes fairly marked. The environmental trend does not linearly follow the axis and it is usually more exact to delimit 'zones' of the ordination which relate to northern/low water availability rather than talk in terms of an *axis* which relates to a north–south effect. In addition species are rarely related exactly and linearly to *one* environmental parameter and thus the *mathematical model, component analysis,* is only an approximation to the real world situation. It is fortunate that these limitations do not detract to any great extent from its considerable power in this area of ecology.

GENERAL DISCUSSION: CLASSIFICATION OR ORDINATION?

In general, the methods of approach to the handling of vegetational units fall between two schools that have developed, apparently representing two extremes. Braun-Blanquet erects a series of units, or associations, which can be classified in a hierarchical system, and are distinct, recognizable and definable entities. Curtis and his associates on the other hand regard vegetation as a continuum, no two stands being the same. These two schools of thought are a direct result of the sampling methods employed, non-random on the one hand and random (or approximately so) on the other. Thus the careful choice of stand with the stipulations of minimal area and homogeneity is highly subjective, and initially at least, only the central 'nodum' (see Poore 1955–6) being isolated, characterized and described. The numerous intermediates are ignored. At a later stage the more obvious intermediates can possibly also be considered, and described as 'associations' or 'facies'. It is debatable whether this ultimate stage is possible in practice. Such an ultimate stage in fact, represents a continuum in the sense of Curtis *et al.* where a random selection of sites necessarily includes 'noda' and intermediates alike. The continuum approach does not produce anything which can be classified, the ordination shows instead the inter-relationships between the species which are believed to be controlled by environmental gradients. However, having obtained this detailed information from an area of grassland for example, one still talks of a *Sesleria* association, or an *Agrostis/Festuca* association or a *Nardus* association purely for convenience. The presence of a large number of intermediates is certainly accepted but the noda are very useful concepts. Implicit here seems to be the *frequency of occurrence* of a particular range of stand variation; a *Sesleria* association is variable but in an area of limestone grassland it is a very common unit, much more so than the intermediate *Festuca/Sesleria* stands. The two extreme schools of thought based on random or subjective choice of samples in fact offer two useful ecological approaches. Man always seems to want to classify and an orderly arrangement of vegetational units is certainly a desirable and useful objective. Conversely it is equally desirable to study in detail gradients of an environment and their effect on species composition. Both are equally valuable.

It is informative to consider at this point which features of a stand of vegetation make a visual impact on the observer. It is not merely the presence of a number of species which form the basis for description but their density, distribution, size, growth-form and their vigour. Thus ideally any classificatory system should be based on the species present, and also on measures of their abundance, distribution, size and vigour. The samples should be taken randomly, relative to the vegetational 'type' to be studied, i.e. 'woodland', 'grassland', etc. and the ordination or classification should be based on an objective method. On this basis the Braun-Blanquet approach falls short of the ideal. It is based on non-random sampling, it employs the measure of presence or absence only to characterize the association, and although some estimate of abundance is available from the description of the stands in the field, little or no subsequent use appears to be made of it.

The use of χ^2 contingency table and the subsequent arrangement of a system of units, either hierarchical or spatial, similarly falls short of the ideal. The method employed by Williams and Lambert picks out species groupings which are given artificial boundaries since at each stage of the procedure the quadrats are repeatedly divided into those with and those without a certain species. This presents a tidy and organized classification but one which is highly artificial and has little ecological meaning. The method undoubtedly arose as a result of the parallel hierarchical classification used in taxonomy with the implied phylogeny and genetical relationships between the different groups. However, in ecology no such relationship exists and a hierarchical classification may well separate two groups of plants which prefer acid conditions since one grows in a wet bog and the other in a dry heath. This is well summarized by Webb (1954), and though such a classification appears to be readily assimilated and orderly, it is an erroneous and over-simplified version of the real state of affairs. The outstanding advantage is that it produces a map of the area studied, with arbitrary lines drawn on the boundaries of the spatial limits of a species or groups of species. Subsequently detailed attention can be directed to those parts of the 'map' which are markedly different and these differences investigated in relation to the local environmental factors.

The final ordination technique discussed, principal component analysis, is by far the most successful technique applied to data ordination yet developed. The concept of noda is clearly still applicable even in the context of an ordination. The population structure in terms of species abundances and/or environmental measures can also be clearly and efficiently defined. The necessity of a computer with a minimum of 1200 K(octal) store does not offer any serious obstacle, and costs for computer time are not impossibly high (plot ordination of 540 samples with 58 species—153·333 sec on a CDC 6400 series). It remains for field ecologists to fully utilize this, and other related techniques, in detailed studies of plant systems so urgently needed in many parts of the world.

11 Digital Computers and Ecology

JUST as computers have had a profound effect on many aspects of chemistry, physics, and engineering, so have they invaded biology and especially ecology. As the complexity of the plant and animal systems examined has been appreciated and as the dynamic or temporal aspects of many ecological situations realized, computers as a research tool have become indispensable. Twelve years ago it was possible to play the 'helpless biologist' in the face of this vastly complicated electronic equipment and welter of jargon. Now in 1973, it is deemed not merely useful to have a grasp of FORTRAN, but also a simulation language and an understanding of *how* the computer installation operates. Knowledge of a systems language relevant to the particular computer installation the user has access to is also strongly recommended. It is not the intention here to attempt a primer course in programming but to try to explain some of the extensive computer jargon and relate it where possible to biological situations. In addition, it is useful to examine the facilities of a computer and indicate some of the areas where the application of computers other than as number-crunchers is particularly useful in ecology. Above all else, the biologist who has a weak mathematical background should not shy away from computers on the mistaken assumption that a deep understanding of maths is necessary.

THE COMPUTER

A computer will only do *exactly* what it is instructed to do; no more, no less.

Since a computer operates on the principle that there is either an electrical impulse at a certain point or there is not—i.e. a 0 or 1 or binary situation both instructions and answers have to be, and are, in this form. This *machine language* usually consists of very unwieldy binary instructions which are difficult for the non-specialist to handle. Conveniently, the computer itself converts simple coded instructions which are user-orientated, into the actual *machine language*:

Thus a simple computer language which we can readily learn can be used to write a series of *exact* instructions to the computer. This *source* program as a deck of punched cards or length of paper tape is then read in and converted automatically by part of the *computer* termed a compiler into an *object program* which the computer can then use as operating instructions. A compiler is an extensive dictionary the computer uses, and each language has its own compiler. The diagram above is a simple *flow diagram* showing the sequential pathway from the user, through the computer and back to the user as a set of answers. It is important to recognize that each arrow and each box represents a discrete step or action which will require varying time periods. The total time the computer is actually in operation is a fraction of the time usually taken up by reading in a program and writing out the answers. Whilst these relatively slow operations are being carried out the most expensive piece of the *hardware* the computer itself, is sitting idle:

The constraints of readers and printers can be overcome by utilizing a number of readers and printers operating in parallel. The object programmes are loaded as a sequence of separate jobs to create an *input file* which effectively queues jobs for the computer. The output is similarly set up as an *output file*, queuing the separate jobs for the printers. Such a system is fairly efficient and ensures continuous operations of the computer. In third generation computers the central memory with its own *central processor* is controlled by a series of *peripheral processor* units which in turn also control the other related *hardware—printers, readers, plotters, disc units, tape units*, etc.

Memory involves a specific number of *hardware* items: *Core memory* the central storage area and often termed *direct access store* which can be accessed randomly. This store is finite and is amplified by the *backing store* composed of memory discs with reading and writing heads on arms which move back and forth over the surface of the disc picking up or relaying information. Discs can thus also be accessed randomly. The backing store usually includes, in addition, a number of magnetic *tape units* which can only be *accessed sequentially*. Both core and disc have the important property that their locations may be *addressed*, that is to say specific items named by the program are located in very specific locations in memory—they have an *address*. The exact address is physically achieved by the computer itself. The name of the address is arbitrarily chosen by the programmer and then subsequently that address name is unique to a specific physical location in memory, unless its uniqueness is removed by further program instructions. Thus DATA (50) is a convenient name for 50 values of plant density, each of which is located in a unique address DATA, and each of which can be found and used in subsequent operations in the program.

The mechanism of operations and the programming of them is beyond our present scope and has been very adequately covered by Maurer (1968), McCracken (1965), Golden (1965), etc. In essence, and fundamental to programming languages are a number of basic *algorithms*. These allow the user to instruct the computer to for example, add, subtract, multiply, divide, square, take the square root, sine, cosine or tangent of an angle, or take the choice of several possible pathways in the program.

The fundamental step in all computer languages is the instruction:

$$X = Y$$

which means literally, replace what is in location X in memory with what is in Y. All the basic operations in a computer are done in this way. Thus, SUM = SUM + DATA (1) replaces what was in the address SUM (i.e. a specific location in store) with the original contents of SUM *plus* the last item of DATA. Sequentially, the sum of the data, for example, can be built up in this way and, in fact, at the machine level, all the algorithms are achieved using this basic step.

Specific languages have been designed for specific computers and some even, with specific purposes in mind. The general programming languages FORTRAN and ALGOL, however, all share the above basic requirements.

COMPUTER USAGE AND FLOW DIAGRAMS

The uses of computers in biology range from a simple utilization of their very rapid arithmetical operations, through the maximum usage of their large storage capacity at the same time as their rapid speed of calculation, to usage of their speed and memory together with their facility to handle additional peripheral hardware. Thus complex calculations can be achieved rapidly. Complex biological systems involving tedious and/or difficult calculations can be analysed and simulated. Voluminous *data files* (simply a sequential list of data results usually on tape) can be summarized and above all graphically displayed using a plotter unit under the control of the computer. The choice of whether to computerize or not a particular basic operation is often a function of the availability of standard *library routines*. A large number of standard programs already exist with specific usage instructions as a library of programs with each computer installation. Thus correlation coefficients, linear regression, multiple regression, and analysis of variance are all standard statistical techniques, each of which is available as an immediately operational program, involving the user in the minimum of effort. Conversely, a complex analysis, requiring the complete development of a unique program, is a very different situation, requiring considerable developmental and testing time. The decision then is simply related to expediency.

The first step in the development of a program is the breakdown of the problem into its specific and sequentially related steps. This is a completely essential step in most program development and is synthesized into a *flow diagram*. Each operational step is symbolized and the pathway of calculation indicated:

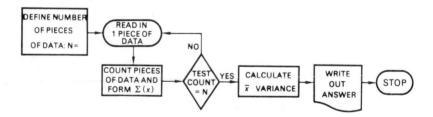

In this simple routine a number of data cards are read in and counted. The sum of x is formed during the reading of the data and finally the mean is calculated and written out as the answer. This flow diagram illustrates the number of steps required to calculate a mean value x. Firstly, (N) is defined and a data card read in and added into a sum (x) location. Subsequent cards are also added into this location so that it finally contains $\sum x$. The count enables the computer to check when it has read in *all* the cards by comparing its own count with the definition of (N) initially given to it. When and only when all the cards have been read in will the next arithmetical step be taken and the answers written out. Thus in addition to showing the sequence of operations this flow diagram also shows how control of the sequential *direction* can be manipulated. Several notations have now been standardized to represent particular processes:

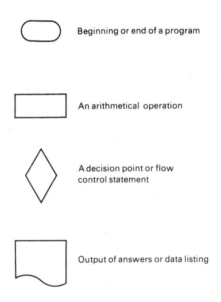

Beginning or end of a program

An arithmetical operation

A decision point or flow control statement

Output of answers or data listing

A further example of flow control is given below (Fig. 11.1) in the context of one possible approach to simulating interference between different individuals of a plant population growing at high density (see Chapter 6).

Working from an initial population of seedlings (N), with a defined spatial distribution and weight, and utilizing leaf area, index/net assimilation and also leaf area index/interference (LIMIT) relationships, over a time period TIME, the individual or population 'growth' can be followed. The changes in sequential direction are controlled by *time of day* which then allows increments of (positive) units of carbon fixed, or increments of (negative) units of carbon respired (Fig. 11.1). Each of the N individuals

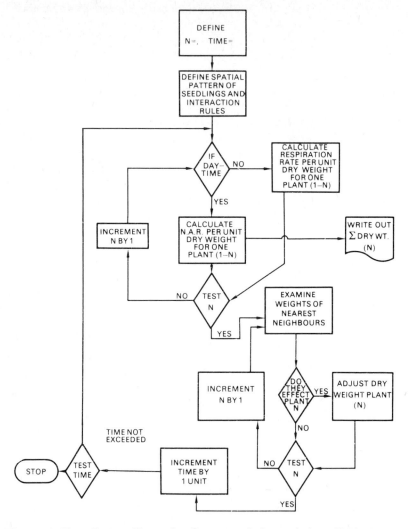

Fig. 11.1. Flow diagram illustrating flow control. A population of N individual plants increments Carbon-units, and each individual interacts subsequently with its nearest neighbour. The whole model is run over a defined TIME sequence.

in the model is grown in this way. The dry weights of the nearest neighbours to each individual are scanned and using the defined relationship between leaf area index/interference, the dry weight increment is suitably adjusted where interference is operating in a positive sense. Again N is tested. Finally each 'pass' through the event sequence on an hourly basis is recorded and tested against TIME in months. Clearly many additional parameters can be incorporated into the flow diagram. In fact, this particular set of relationships has been summarized as 'Growth and inter-action' in the diagrammatic systems representation used above (Chapter 6, pp. 104–127). Conversely, the examination of nearest neighbour dry weights for assessment of interference between individuals, necessitates its own flow diagram of the two-dimensional search routine required. Thus a complete flow diagram is often developed as sections or even subsections. Ultimately each block of the diagram is broken down into a final series of programme instructions to the computer.

The development of a flow diagram helps considerably in simplifying the whole organization of the program and above all aids in defining those parameters which will require changing if the final program is to be generalized. It is only in very unusual situations that a program is worth writing when it is to be used only once. The majority of pro-grams become standard routines which are used frequently by one or many individuals depending on its breadth of application. It is important therefore that the, design of a program be related to general usage, with the number of samples, the number of replicates, the number of species, etc., given by algebraic notation. At the beginning of the program these controlling parameters can then be defined in relation to the specific set of data being analysed. The flow diagram assists considerably in ensuring all such parameters are covered. Thus inherent to a good flow diagram is not only the concepts of dynamic sequencing of operations and the necessary directional control, but also an optimal usage of the computer from both the point of view of *program efficiency* as well as *user requirements*.

DATA HANDLING AND DISPLAY GRAPHICS:

In any situation where the use of a computer is being considered for a series of trivial but numerous calculations and where there is some hesitation over the amount of time ultimately saved by computerization, two other aspects should always be considered: obtaining field data in-volves a considerable amount of effort and over a period of time is often re-utilized in new approaches to a topic. A permanent *data file*, as punched cards, tape, or magnetic tape plus a data tabulation as a printed record, is not only extremely useful to the individual but can also be subsequently made readily available to a number of other interested investigators. The necessary program development in this situation is short and the advantages considerable. Again the speed of the computer is being utilized together with its linkage to a high-speed printer, saving con-siderable effort in producing a 'fair copy' of the original data. The time required to transfer the field data to cards or tape may be lengthy, but it represents also a permanent record of some considerable potential and plasticity.

A further consideration is the actual output form of the results of a data analysis. Very often a series of analyses results in complex tables of numbers which are then subsequently hand drawn into graph form for rapid visual appreciation and summarization. In many instances this can be extremely time consuming work which the computer can also be programmed to execute very effectively. Just as a program language can control arithmetical or logical operations in a computer, then similarly certain program instructions can be used to generate a magnetic tape file. This then serves as the operational control of a high-speed *plotter* which is a separate but an integral part of the peripheral hardware of any computer installation. The magnetic tape contains coded instructions, generated by the computer, which control the movement of the plotter pen. There are only six instructions.

1. Raise Pen
2. Lower Pen
3. Move pen 1 increment in a +X direction
4. Move pen 1 increment in a −X direction
5. Move pen 1 increment in a +Y direction
6. Move pen 1 increment in a −Y direction.

An instruction in the X and Y direction can be given simultaneously and the pen can thus move in 8 directions, $45°$ apart in the two dimensional X–Y plane. Using extremely small increments ($\leqslant 0.005$ in) even a continuous curve, although made up of a series of line segments, appears as a smooth curve to the naked eye. Present-day plotting units produce a plot of a professional draughtsman's standards of accuracy, but extremely rapidly.

Two examples will suffice to demonstrate the power of this particular approach: Kershaw and Rouse (1971) have shown that a ground cover of the lichen *Cladonia alpestris* very largely controls the soil water characteristics of the underlying substratum during the summer months. It is probable that the lichen cuts off incoming energy to the soil surface thus removing the source of energy for the evaporation of soil water. This situation equally should be reflected in the soil temperature profile. An investigation of soil temperature profiles under different lichen surfaces, with 14 replicate measures of temperature at the bottom of the lichen mat and with soil temperature profile data at 0, 5, 10, 15 and 30-cm depths produces a considerable quantity of data. The flow diagram to handle the data and plot the output is given below (Fig. 11.2) with an example of the output from the computer and a Benson-Lehner plotter (Fig. 11.3). Thus in addition to the arithmetical operations for each 15-minute period and

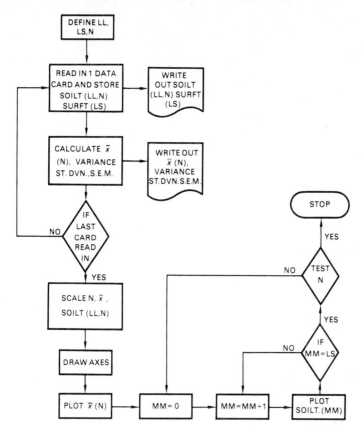

Fig. 11.2. Flow diagram of the program to calculate means, variance, standard deviation and standard error of the mean, for (LL) soil temperature profile values and (LS) surface temperatures (SURFT) measured every 15 minutes over a two-day period, and to plot the o/p.

a file listing of the raw data, the graphical output enables an immediate data appreciation to be made. Since soil temperature profiles under two lichen species were examined over a 60-day period, with readings taken every 15 minutes, this represents a considerable saving of effort. The program is a general program and extremely simple, with LL, LS and N defining the number of probes used at the bottom of a lichen mat, the number of levels examined in the soil profile and the total number of data cards in one run respectively. The program can thus be used for a variety of data, covering all situations likely to be met within this field.

A second example is taken from the use of component analysis as an efficient ordination technique (see Chapter 10). The computer is an essential requirement for the calculation, and the resultant output is usually

Fig. 11.3. An example of the plotted temperature profile o/p.

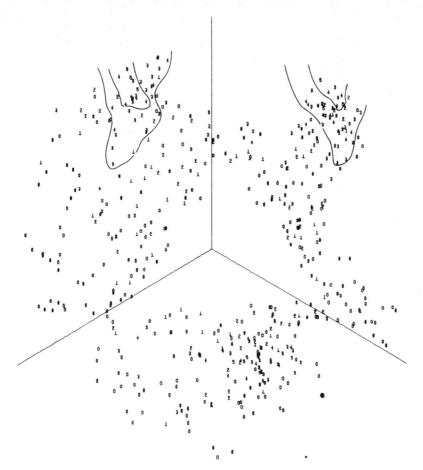

Fig. 11.4. (A) Species abundance (Domin values) overlay diagram for interpretation of component analysis ordinations as computer graphical o/p for *Alectoria nitidula*. The contour lines were hand-fitted to the computer graphics output.

examined as a projection of the constellation of points, each representing a single sample, on the plane surfaces formed by pairs of axes (cf. Figs. 10.17 and 10.18, pp. 220–21). Subsequently values of abundance for selected species or environmental parameters can be overlayed on the ordination diagram and an interpretation of the different axes made. Again for ordinations of several hundred plots, with overlays of the more abundant species, several weeks of extremely tedious graphical work are involved. The projection of the points on to the three plane surfaces represents, in fact, only an approximation to a three-dimensional model. The draughtmanship involved in drawing such a three-dimensional model is considerable and further complicated by the frequent shielding

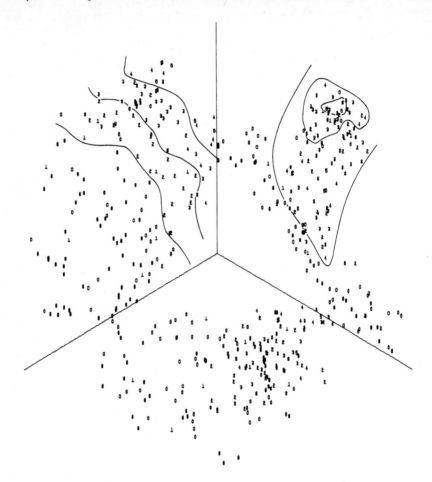

Fig. 11.4. (B) Species abundance (Domin values) overlay diagrams for interpretation of component analysis ordinations, as computer graphical o/p for *Cladonia arbuscula*. The contour lines were hand-fitted to the computer graphics output.

of plot symbols in the 'background' by those in the 'foreground'. Hence the need for simplification by projecting on to each plane surface and considering the spatial inter-relationships in successive pairs of dimensions. Kershaw and Shepherd (1972) have made an approach to this problem by using a rotational graphics model enabling a 'view' of the ordination to be obtained from any defined angle or orientation desired (see Fig. 10.21, p. 226). This is tied to the traditional projection on the plane surfaces of any (stipulated) dimensional trio, with overlay diagrams of any number of (stipulated) species or environmental parameters. An example of part of the graphical output is shown in Fig. 11.4 for two of the species abundant in the lichen heath samples from northern Ontario (cf. Fig. 10.21). The amount of time saved is in this instance very considerable for the plane projections and species overlays alone, which are normally hand drawn. The rotational model representations is purely a computer product and would not be possible at all without the facility of the computer and the associated hardware. The development of the program is clearly complicated and lengthy. However, ordination techniques are a standard multivariate method, now widely employed in both vegetational analysis and taxonomic situations. Component analysis will be used dozens of times in a series of analyses over a period of months, and the saving of time and labour by use of the graphical output very quickly outweighs the time spent in the development stages. The program is generalized so that numerous other users can readily adapt it to their own particular requirements leading to an even more extensive optimization of the overall usage of the developmental time.

Not only is the computer a number-cruncher of impressive proportions, it has a number of other attributes which have not as yet been widely used in biology. Simulation studies are one such use (see Chapter 12), other possibilities are only briefly indicated here. Many more will be appearing in biological journals.

12 Computer Simulation Studies

ALTHOUGH plant ecologists are aware of the fundamental importance of time as an ecological parameter, and that it is one of the central concepts in succession and cyclic phenomena, it is usually avoided as an experimental vector. Where changes in plant populations have been examined, the experimental material has been chosen carefully from annual fast growing species, or only changes monitored during the first few months of the population growth. Equally, although there is an awareness that a plant association is controlled by several or numerous environmental parameters, plant ecologists still talk of single factor control. Thus 'competition effect', 'pH effect', 'soil depth effect', etc., are all widely used descriptive (although loosely defined) terms. Embodied in this attitude for example, has been the approach to an insect crop pest—spray it with DDT.

The reasons are straightforward. To experimentally, but exactly, control a population of plants over a ten-year period is not practical. To consider simultaneously more than two parameters, which interact together and control the growth of a plant, is extremely difficult. Thus although the concept of an ecosystem is long-standing, the approaches to the analysis of its structure and dynamics have been very limited. More recently, the advent of computers has given a tremendous stimulus to the study of *systems* as complex and changing entities. This so-called systems approach is not a new *concept* to ecology, it is a more sophisticated and realistic *approach*. By integrating many of the already existing ecological methods, techniques, and principles, utilizing the powerful facilities of a present-generation computer, a synthesis of a complex ecological situation can be modelled.

GENERAL PRINCIPLES

Among the earliest attempts at a holistic approach were those of Odum (1956), Odum (1957), and Odum (1960), which were attempts at analysis of energy flow through systems. Their approach resulted in a flow diagram of productivity (Fig. 12.1), which has been widely followed by other

Fig. 12.1. Energy flow diagram with estimates of energy flows in kilo-calories per square metre per year in the Silver Springs community. (From Odum 1957; courtesy of *Ecol. Monogr.*)

ecologists (see also Odum 1971). The historical impact of the models was considerable and focused attention on one important aspect of the eco-system—energy relationships between the components. However, lacking in their approach is any concept of time control of the biological mecha-nisms or intrinsic control by the individual of its utilization of the external energy sources. Hubbell (1971) very rightly questions the sterile nature of much of the work on productivity and energy flow due to the failure to recognize the existence of control mechanisms over the use of available energy: 'In my view the prevalent treatment of organisms as passive agents has hindered further development in the field of ecological bio-energetics by producing few significant questions about what living systems are really doing with energy. Such a treatment ignores perhaps the most fundamental characteristic of life: The capacity of living organisms to regulate, within the bounds established by the laws of thermodynamics, the rates at which they accumulate and dissipate energy.' The study of such controlled utilization of energy in a system would be of considerable interest.

A computer is an extremely fast 'book-keeper' which can keep track of a very large number of profit and loss accounts, items in stock, on order, and with some additional analysis, can predict on the basis of the last three years sales and current trends, next year's requirements. It is exactly these facilities which can 'book-keep' the dynamics of an assemblage of individuals, the growth of a species over any (defined) time span, or even the physiological processes involved over short time periods:

This simple simulation model, although an extreme reduction of a much more complicated system, illustrates many of the fundamental aspects of computer simulation (deliberately, only a single environmental variable is used here for simplicity). The model simply advances through a time sequence N, until it reaches the end of the stipulated period LIMIT, 'fixing' carbon units at a rate determined by a variable parameter LIGHT, and a defined relationship between light and the amount of carbon fixed. The following criteria are fundamental, and also common to other computer simulation models:

1. This model relates to a particular level of organization—the cellular level. It would be equally possible to work at the sub-cellular level and model reaction rates of enzyme systems, at the level of the individual and increment dry weight increases, or at the association level and increment biomass through each time increment. Thus the initial decision necessary in the construction of a model is the actual level of detail necessary for a solution of the problem in hand.

2. The model involves relationships which have been experimentally determined, prior to the formulation of the model. Thus the relationship between light intensity and rate of C-fixation must be known independently.

3. The time sequence $N = N + 1$ in the model above is incrementing arbitrary time-units. The actual units used, seconds, minutes, hours, or days, also necessitates a decision. This is usually based on the maximum length of simulated time (LIMIT), the degree of accuracy required in the model, and the nature of the variable parameters. Thus if the carbon-fixation is being simulated over a period of 24 hours with the light varying both diurnally and with cloud cover, time increments should be in minutes. Conversely over periods of a year, 15-minute time steps would be more appropriate. The size of the time interval and also the method adopted to determine the carbon increment during time N to time $N + 1$, has considerable bearing on the level of accuracy achieved by the model. If the time increment is relatively large, integration will be necessary, small time increments effectively achieving the same purpose.

4. A simulation model is a deliberate simplification of a real system, all unnecessary information being discarded. An exact copy, if it were possible, would be as bewilderingly complex as the real world and thus of more limited value. The C-fixation model above, although very incomplete as it stands, would never incorporate changes in soil water nutrient concentration, for example. This is an unnecessary complication in this particular model. It is only fundamental to a model operating over longer time scales and involving simulation of plant *growth*, controlled by a much larger number of variables.

5. The organisms involved in the model have some order of internal control in response to external changes during the time sequence used. The simple C-fixation model does not have this feature—the carbon pool would go on filling up until the 'cell' burst. A more realistic, yet still very simple model would be:

The 'cell' now has a simple mechanism of control over the size of its carbohydrate pool. This model is a more exact representation of the normal situation where overall cell metabolism rates are often dependent on the removal or conversion rate of the final product or of one of the intermediaries of the synthesis pathway. There is *negative feedback* in the model giving a simple internal level of control over the utilization of the external variable, light. Equally positive feedback is a fundamental mechanism in many systems models.

6. Because of the complexity of even simple biological systems, models are constructed as a series of block-units, initial efforts being directed at strictly defined (small) parts of a system which subsequently can be related to other adjacent block-units progressively encompassing a larger part of the 'total' system. At each step the model is tested against independent observations on the real system. The illustrative C-fixation model used above is representative of a single initial block-unit.

There are several other features of biological systems which are equally fundamental to simulation models but which are not represented in our C-fixation model and are best considered as case-studies (see below). Hubbell (1971) states: 'Biological systems particularly are full of threshold phenomena and saturating nonlinearities (Holling 1966), synergisms (multiplicative effects), hysteresis, and the like.' Thus the initial approach to the simulation of a system is to develop a simple linear model. That is to assume that if $Y_1 = f(x_1)$ and $Y_2 = f(x_2)$ then $(Y_1 + Y_2) = f(x_1 + x_2)$ or, that there is no second-order interaction between the two variables controlling a process. *Often this is not true*, and a linear model may only give an approximation (reasonable or otherwise) to the real system (see below). Inherent in biological systems are non-linearities. This again emphasizes the importance of thoroughly testing the model as it develops so that non-linear segments may be incorporated if required.

Similarly hysteresis effects—the lagging behind of what should be a normal level of response, after the system has been under a period of stress, can induce serious inaccuracies in simple simulation models (see case study 4 below).

It is of particular significance that the structure of several computer languages are ideally suited to the handling of ecological systems. The concept that a complex system can be broken down into a number of 'discrete' building blocks, which subsequently can be resynthesized into the total system, parallels closely the algorithm concepts of FORTRAN and ALGOL. The iterative DO-loops, IF, and GO to control statements are similarly very powerful computer language functions, in setting up sequential time flow, threshold limits, interactions, feedbacks, etc. The choice of a computer language to some extent determines the strategy employed in a simulation study. Other considerations are involved, and before we consider a number of case studies, it is necessary to examine some of the overall strategies that have been employed.

GENERAL STRATEGIES

1. ANALOG STUDIES

H. T. Odum (1960) used an electrical network to mimic the flow of carbon through an aquatic system, by a flow of current. The flow of current was adjusted by variable resistors to equate with turnover rates obtained from field data. Neel and Olson (1962) point out the extreme limitations of electrical analog circuits and that such net-works are devoid of time-dependent components, which are so characteristic of ecological systems. They also point out that electric network analog solutions are particularly prone to errors introduced by the components themselves, counter-e.m.f. errors due to capacitor components, and the measuring devices introduced to monitor current or voltage reducing the 'true' potential differences across the circuit components.

Conversely electronic analogs do not have these limitations to anywhere near the same extent and can be used very effectively for some simulation models, and in particular compartment productivity models (see below). The basic arithmetic processes of addition, subtraction, multiplication, and division are readily achieved and, above all else, a very rapid and simple integration of any desired function. The fundamental electronic component is a high gain operational amplifier. If a number of signals at arbitrary potentials are applied to the control grid of the operational amplifier the O/P signal will then be the algebraic sum of these potentials. The magnitude of the O/P can be simply controlled by the gain of the amplifier. Since sign and phase inversion occur in each amplifier, summation of 'out-of-phase' signals effectively represents subtraction. Thus if a voltage is applied which fluctuates in proportion to photosynthetic rate, a very simple combination of electronic components will simulate carbon-fixation. Using the notation of Olson (1964) (Fig. 12.2) our original C-fixation model simply reduces to:

Analog circuits can reduce block segments of a simulation model to a few electronic components. Olson (1964) gives a typical compartment

Fig. 12.2. Relations between productivity rates (a), (b), (c) and integrated photosynthesis (d), assimilation (e), and plant accumulation (f) simulated by analog computer components. O/p of amplifier (at lower point of triangle) = sum of i/p voltages with sign changed in (a), (b) and (c); o/p amplifier = integral of this sum if an accumulator is attached [box in (d), (e) and (f)]. (From Olson 1964; courtesy of *J. Ecol.*)

productivity model and its analog representation which can be quickly set up (Fig. 12.3). In practice this is achieved simply by 'plugging in' the required components in the correct sequence with 'plugs' through some type of control board analogous to a telephone switch board. This is usually removable so that the network can be set up remote from the computer ready for use, and although simple models will only require an elementary knowledge of electronics, more complex models with non-linear segments and hysteresis effects require a much more active grasp of the electronic methods now available (see Patten 1971).

Although analog computers can be particularly powerful in the field of energy flow through an ecosystem using the compartment model approach there are a number of limitations. Availability of analog computers is restricted and, in addition, accessibility may be strictly limited—there is not the same structured operation as in a digital computer (see chapter above) and hence facility of use. Above all else analog methods do not

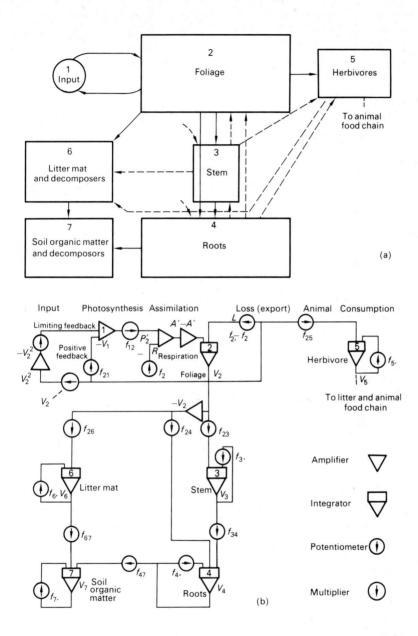

Fig. 12.3. Model ecosystem as a compartment model with the corresponding analog circuit. (From Olson 1964; courtesy of *J. Ecol.*)

have the powerful advantage of logical decisions and pathway choices as readily available as in a digital computer, and can thus not easily be made to scan a list of possible options, an essential part of many simulation studies.

2. DIGITAL STRATEGIES

The majority of simulation studies have employed standard digital computer methods and often utilize the standard languages of FORTRAN or ALGOL. There are considerable advantages to the use of these languages for simulation purposes—there are a large number of standard library sub-routines of curve-fitting, integration methods, statistical analyses, and function generators, already available. In addition FORTRAN compilers (especially) are available for all present-day computers. The same is not true for many other languages designed for various simulation situations, and which have time sequencing of events 'built into' the language. Both Chapas (1969) and Radford (1969) make out a case for DYNAMO as a strong simulation language for biological situations. It probably is, but most computers do not compile it. Compatibility with other computer installations and hence the *generality of the model in a computer sense* is an important concept.

Simulation models may be *stochastic* or *deterministic*. This simply denotes whether a model contains some random events or not.

If a sequence of events follows with complete certainty then the model will produce a result not affected by chance events: it will be deterministic. Conversely if one or more of the sequence pathways is selected on a probability basis, the model is stochastic or probabilistic.

Mathematically, random events are more difficult to handle and as a result most models deal with 'an average' situation in a deterministic model, which can then be handled by the powerful mathematical techniques which are available. Random events in biological situations, however, are widespread. In the sense of a simulation model an 'average' *individual* will adequately describe the response of that average individual to the system, but conversely, if a model *population* is involved, random events may be necessary to describe adequately the structure of an 'average' population (see case study 3 below).

The mathematical approach to a computer model defines a number of *system variables* which characterize and control the 'flow' between component segments of the model. The flow or interaction between the components are derived from *transfer functions*. The model *driving equations* (*forcing functions*) involve those external parameters required as input to the model but which are little affected by the model if at all. Usually the system is represented by a series of differential equations and 'solving' the model equations means determining the value of the variable V_i over time from a given initial value. The range of mathematical techniques involved are outside the scope of this book, but see Patten (1971).

Two basic strategies have arisen: *Compartment models*, which are usually simple linear systems of differential equations very often describing biomass changes or energy flow in a system. Their purpose is to analyse the performance of the model rather than necessarily to attempt to model the real system. The control of 'steady state', and the relative degree of control of different component segments over the total system, can be evaluated this way. Thus a system is represented as a series of compartments, or pools, of energy or nutrients. The complex interactions *within* each *compartment* are assumed to counter-balance one another to some extent and the overall behaviour of each individual compartment is thus simplified. The basic data on which the model is constructed is usually a gross estimate of 'compartment size' in terms of the biomass of roots, stems, leaves, etc., and the whole approach can suffer from the unrealism of the basic philosophies involved. The mathematics can be elegant, but the structure of the model often lacks several of the 'control' features fundamental to biological systems (see above). The value of the approach lies in the fact that the model is simulating the production of a complex *population* of many species. An approach to such a system modelled at a more detailed level is impossible with the present lack of ecological information.

The second and more detailed strategy that has been employed is the component approach of Holling (1966) or building block approach (Kershaw and Harris 1969). This approach, incorporating all the general criteria outlined above, proceeds in a stepwise fashion, from an experimental examination of each block or component, in turn, to a series of equations relating the parameters involved in each component, and back to the experimental-testing of the model at each step. This approach has so far been largely restricted to processes operating at the level of the individual organism, and the behaviour of a population is inferred from the individual's response. Despite this present limitation the component approach has two outstanding attributes: (1) the stringent requirements of a computer model enforces an examination of the real world system in a very different way, often resulting in a new and better understanding of long-accepted relationships; (2) there is considerable 'feedback' from the model even when it is in an early stage of development.

The following case studies are illustrative of the component approach to computer simulation and of the basic criteria involved in model construction.

CASE STUDIES

Case Study 1—The Component Approach

McQueen (1971) presents an elegant experimental situation involving two species of cellular slime moulds. Their life cycle runs from spore to vegetative amoebae, through aggregation, and fruiting body production, back to a spore. The component flow diagram (Fig. 12.4) shows this life

Fig. 12.4. Flow diagram of the components of the finished model (see text). (From McQueen 1971; reproduced by permission of the National Research Council of Canada from the *Can. J. Zool.*, **49**, pp. 1163–1177.)

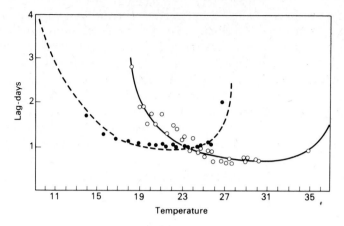

Fig. 12.5. Spore germination lag plotted against temperature (°C); ●, *D. discoideum*, ○, *P. pallidum*. (From McQueen 1971; courtesy of *Can. J. Zool.*)

Fig. 12.6. Growth of amoeba colony area against time for *D. discoideum* (●), and *P. pallidum* (○). (From McQueen 1971; courtesy of *Can. J. Zool.*)

cycle for two species and their interactions with the limit imposed by food resources and space availability. For both species, from experimental observations, the lag time between germination and amoebae production decreased as temperature increased, reached a minimum at an optimum temperature, and increased with further temperature increase. The equation to fit these observations is based on three assumptions: (1) spores do not germinate below some temperature T_L; (2) spores do not germinate above temperature T_H; (3) germination time is minimized at T_0 which lies between T_L and T_H. The resulting equation fitted to observed data (Fig. 12.5) accounts for 75 per cent and 88 per cent of the variation in *D. discoideum* and *P. pallidum* respectively, and is the first subcomponent of the model.

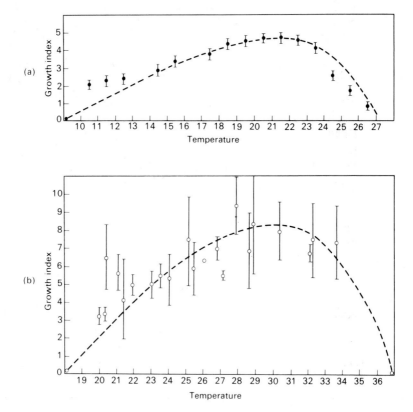

Fig. 12.7. (a) Observed (●) and predicted (– – –) colony growth of *D. discoideum* against temperature (°C). (b) Observed (○) and predicted (– – –) colony growth of *P. pallidum* against temperature (°C). Data points are means ±95 per cent confidence limits. (From McQueen 1971; courtesy of *Can. J. Zool.*)

The expansion in area of the colony in both species followed a linear sequence when the square root of the area was plotted against time (Fig. 12.6).

Thus for both species

$$A_t = \{1\cdot7728 \cdot g[t - L(T)] + C\}^2$$

i.e. area at time $t = f$ (time, lag time and constants $g + C$). In addition temperature interacts with rate of growth and an equation to describe the observations was developed to meet three conditions: (1) the rate of expansion at low temperature T_L was zero, (2) the rate of expansion at some high temperature T_H was zero, (3) the maximum rate of colony expansion was at T_0 between T_H and T_L. The observed/predicted results (Fig. 12.7) for the two species accounts for 96 per cent and 56 per cent of the variation for *D. discoideum* and *P. pallidum* respectively. The marked scatter of data in the latter species was due to bacterial contamination of the experimental cultures.

As the amoebae use up food and space those in the centre of the colony with dwindling resources secrete 'acrasin', which initiates aggregation and fruiting body production. There is again a temperature-dependent lag time, and the same equation developed previously for the lag time of spore

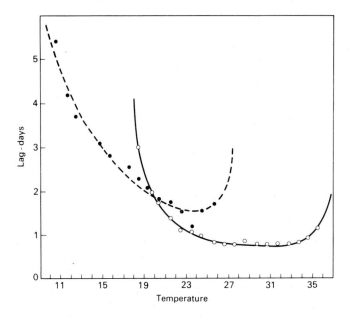

Fig. 12.8. Observed and predicted fruiting body lag (days) against temperature (°C) (●), (− − −), for *D. discoideum*; (○), (——), for *P. pallidum*. (From McQueen 1971; courtesy of *Can. J. Zool.*)

germination gives a very good fit to the experimental data for lag time and fruiting body formation (Fig. 12.8). The process of fruiting body expansion similarly follows the relationship developed for colony expansion, the temperature dependent rate for *D. discoideum* fitting the rate for colony expansion, that of *P. pallidum* requiring a different equation (Figs. 12.9 and 12.10). Finally spores are produced at a known rate and the model run under a food renewed situation until the fruiting bodies cover a maximum limiting area LIMIT (see Fig. 12.4).

Fig. 12.9. Observed (●), and predicted (– – –) fruiting body growth in *D. discoideum* plotted against temperature (°C). Data points are means, ±95 per cent confidence limits. (From McQueen 1971; courtesy of *Can. J. Zool.*)

Fig. 12.10. Observed (○), and predicted (——) fruiting body growth in *P. pallidum* plotted against temperature (°C). Data points are means, ±95 per cent confidence limits. (From McQueen 1971; courtesy of *Can. J. Zool.*)

The simulation of growth of each species was tested against independent experimental data. The data for *D. discoideum* is given below in Fig. 12.11, and shows for the vegetative phase no significant differences between observed and simulated growth at eleven temperatures. The area covered by fruiting bodies shows only one significant departure from simulated area (Fig. 12.12). *P. pallidum* shows very similar results. The model of exploitation of available resources thus satisfactorily predicts both the form and rate of colony expansion for both species over the entire range of temperatures used. The model was further tested by growing the two

Fig. 12.11. Observed (●) and simulated (×), (– – –), amoebae growth for *D. discoideum* over the temperature range 14·5 to 24·5°C. The data points are means ±95 per cent confidence limits.

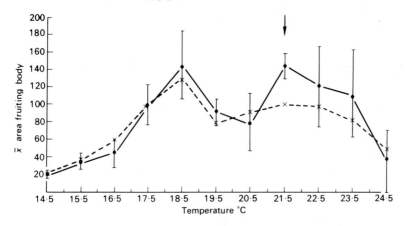

Fig. 12.12. Observed (●) and simulated (×), (– – –), fruiting body growth for *D. discoideum* over the temperature range 14·5–24·5°C. The data points are means ±95 per cent confidence limits.

species together and compared with a simulated o/p derived from the above equations of the simple growth models of each species independently. Food and space used by both species were incremented with two LIMITS to stop colony expansion when all the food and space was used up. The predictions were again tested experimentally and, in short, the model based solely on exploitation does not describe the competitive interaction and presumably the two species directly interfere with one another. The combined model is thus not linear. For additional details of the nature of the interaction and its modelling see McQueen (1971).

This case study demonstrates the utilization of a simplified system with strictly defined limits, the experimental component approach and a typical non-linear system where second-order interactions decide the outcome of competition between the organisms.

Case Study 2—Simulation of Environmental Parameters

The external parameters which drive a model are often meteorological vectors and are readily available from meteorological stations. Thus Harris (1972) used two years' data from Plymouth, in the form of hourly record cards, to drive a model. The cards contained punched information on year, month, day, hour, cloud cover, wet and dry bulb temperatures, vapour pressure of water in air, relative humidity, wind speed, rainfall amounts and duration as well as information on cloud types and the state of the ground. In many instances, however, micro-environmental parameters are required that are not a standard measurement at a weather station and it becomes necessary to simulate these from the parameters that are available.

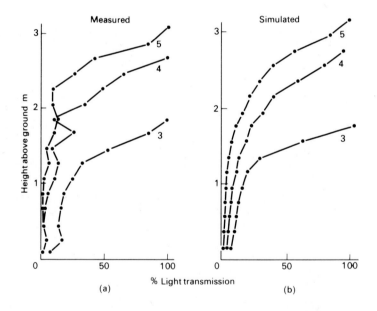

Fig. 12.13. Simulated (a) and measured (b) light penetration at solar noon, into a corn crop at three stages of growth (3, 4 and 5). (From Loomis *et al.* 1967; courtesy of Academic Press.)

Waggoner and Reifsnyder (1968), and Duncan *et al.* (1967), have modelled evaporation and photosynthesis respectively, deriving the required microclimate parameters within the crop canopy from standard meteorological data and the physical characteristics of the crop environment. Thus Loomis *et al.* (1967) give observed and simulated light penetration at solar noon into a corn crop at three stages of growth (Fig. 12.13) showing close agreement. Harris (1972), examining productivity of lichens epiphytic on tree trunks, required to simulate the 'wetting-up' rates of different heights of the trunk. The photosynthetic and respiratory rates of lichens are dependent on their degree of saturation and it is essential to know the rate of arrival of rainfall at any point during time, as well as the subsequent rates of evaporation. The strategy employed in the wetting-up model is interesting.

It was assumed that to some extent, rainfall was influenced in the same way as light. If a portion of the tree canopy intercepted, say, 15 per cent of the incident diffuse light, then it was assumed that this same region would also intercept, though not necessarily retain, 15 per cent of the incident rainfall. This assumption will only hold true for randomly distributed, vertically falling raindrops, and wind will clearly affect considerably the pattern of interception. However, in the preliminary model, aspect effects in relation to the prevailing wind direction were ignored and similarly the effect of wind speed on the model was also ignored. The light profile in the canopy is readily measured using a standard photo cell during summer and winter conditions (Fig. 12.14) and the fall in intensity from top to the bottom of each zone estimated. This value then represents the total 'interference' within each zone and can be used to partition rain and drips within the model canopy (Fig. 12.15).

Fig. 12.14. The decrease of light intensity with height in an oak canopy in (a) winter and (b) summer. (From Harris 1972; courtesy of *J. Ecol.*)

Fig. 12.15. Comparison of the observed light profile with the interference profile used to simulate wetting-up rates. (From Harris 1972; courtesy of *J. Ecol.*)

From experimentally determined data of bark water-holding capacity at different tree heights, six horizontal elements of bark each 1 cm² at each of six height zones represents a constant water reservoir at each height, which is gradually filled with the proportion of incoming rain intercepted at that height. Once it is 'full' further intercepted water then 'drips' on to lower zones and is also partitioned. Zone 1 (i.e. the top sixth of the tree) will intercept 0·12 of the incoming rainfall (from the light profile data). Zone 2 intercepts 0·15 (from the light profile) of the rain *passing through* Zone 1 (or 1·0−0·12=0·88 of the incoming rainfall). Thus Zone 2 will intercept 0·15×0·88=0·132 of the incoming rainfall. This process is repeated at each height in the tree. The distribution of drips is calculated in exactly the same way. Zone 1 receives no drips. Zone 2 intercepts 0·15 of the drips from Zone 1 when the unit area of bark is saturated. Zone 3 intercepts 0·16 of the drips from Zone 2 and so on, each zone receiving drips from all the zones above. The flow diagram of this component of the model is shown in Fig. 12.16.

The comparison between simulated and observed wetting rates under summer conditions (Fig. 12.17) shows very good agreement.

This model segment illustrates very well the derivation of the micro-environmental parameters necessary to drive the net carbon assimilation component of the system. In addition, it illustrates a complex feedback system and the method of programming adopted.

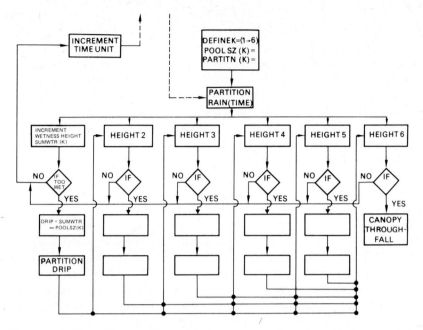

Fig. 12.16. Flow diagram illustrating the strategy adopted to simulate wetting rates in an oak canopy, at six different heights. The incoming rain and drips arriving at any one height zone are partitioned, the residual quantities being available to affect lower levels in the canopy.

Fig. 12.17. Observed and simulated quantities of rain required to saturate sample probes at each height zone under summer conditions. (From Harris 1972; courtesy of *J. Ecol.*)

Case Study 3—Population Simulation

Hardwick (1969) describes a model with built-in random elements to simulate exclusively the development and growth of peas. A pea crop has a commercial value which is a function of the *population* of pods and to simulate events relating to an average pea pod negates the purpose of the simulation study. It is essential to examine how the structure of the total population changes with time and environmental changes.

The sequence of flowering, pod formation and ovule development in peas follows a definite sequence. Two lateral meristems are developed at each node. The first invariably develops into a flower, the second either into a flower or a vegetative process depending on field conditions. Subsequently the apical bud can continue to produce flowers until it dies. Of the flowers that open, flower abscission might occur or not, a pod might develop or a number might abort. These losses from the harvest potential are controlled by a variety of factors which were incorporated into the model. The strategy of the model followed closely the methods standard to C.S.L. (Control and Simulation Language). The podding nodes were a population, each of which had a number of attached pods, peas per pod, and mean weight per pea. During simulated time each podding node was moved into a 'set' during which time a particular series of events took place. At a defined time the podding node moved into another set where further tests took place. The events in each set related to biological events (flower development, or abscission, etc.) and the outcome of each event was simulated by an o/p from the computer random number generator. This number referenced a particular entry of a frequency table of the event 'happening' in that particular set. Thus the number of peas developing in a pod or the number of pods developing at a node is simply generated by a frequency distribution stored as a table. If the random number is 5 then the number of pods developed at that node is the number stored in the 5th location of the frequency table relating to that set. Different nodes move into each set after a lag phase and thus a seasonal effect can be generated in terms of numbers of pods per node by simply having replicate frequency tables with different mean values and standard deviations.

Case Study 4—Hysteresis Effects

Currently there is considerable interest in the rapid eutrophication of lake systems as a function of large nutrient increases. These nutrient increases are a result of man's activities and have the subsequent effect of rapid algal growth. Thus the mechanism of algal blooms, their time sequences, and interaction with the total lake system, are of some concern and importance. Considerable effort has been devoted to measurement of photosynthesis rates of algal populations both in the field and under laboratory conditions. Basically since light intensity falls off with increasing depth in a lake, and since there is continuous mixing of the upper lake levels, it is essential to any predictive model to establish the rate of photosynthesis of the algal population at varying light intensities. The approaches that have been made have all assumed a simple linear relationship which levelled off at some specific point when light saturation of the algal photosynthetic mechanism was reached, with photo-inhibition thereafter. Thus in a steady state system with no turbulent mixing of the upper lake levels, the diurnal light fluctuation would induce a corresponding diurnal pattern of fixed carbon at different depths in the lake (see Vollenweider 1969) (Fig. 12.18). Recently, Harris (1974) working with several algal species has shown that in fact there is a marked

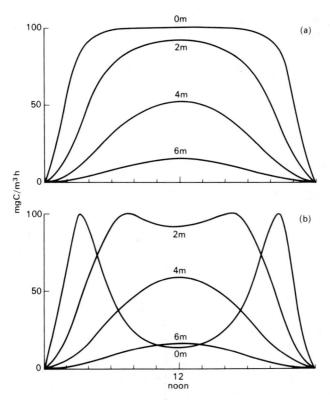

Fig. 12.18. Simulated rate curves at various depths during a standard light day for two assumptions: (A) sub-surface light inhibition negligible, (B) showing strong sub-surface light inhibition. (From Vollenweider 1969; originally published by the University of California Press; reprinted by permission of the Regents of the University of California.)

hysteresis effect (Fig. 12.19). Thus after exposure to light values above 0·4 Langley's per minute, the subsequent response to lower light values follows a typical hysteresis loop (dotted lines in Fig. 12.19) with a markedly different return pathway after exposure to light values of above 0·5 Langley's per minute. The exact mechanism of this effect is not clear, but it is evident that in a steady state system, the symmetry of the diurnal fixed carbon quantities will be considerably modified in the upper levels of the

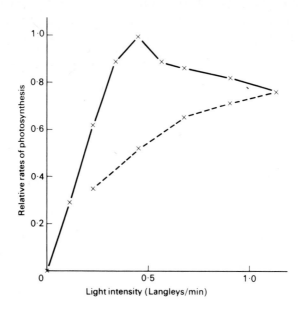

Fig. 12.19. Relative photosynthesis rate in a diatom population (mainly *Asterionella*) exposed to different light intensities (Langleys/minute). The return pathway after a maximum of 1·2 Langleys/min., × – – – × shows a marked hysteresis effect following the light stress period.

lake. The simulated curves are given below together with the equivalent observed data for *Cosmarium botrytis* at a light régime equivalent to one-metre depth showing clearly the extent of the differences in the top three metres (Fig. 12.20A and B). Since the upper levels of a lake are not in a steady state but subject to varying rates of turbulent mixing, the total algal population will be affected for varying periods of time. The initial assumptions of a simple linear rate response up to a fixed light saturation value clearly leads to simulation values markedly at variance with the existing situation.

The strategy adopted to simulate carbon assimilation is as follows: The incoming energy input RZ at any TIME period is used to calculate the

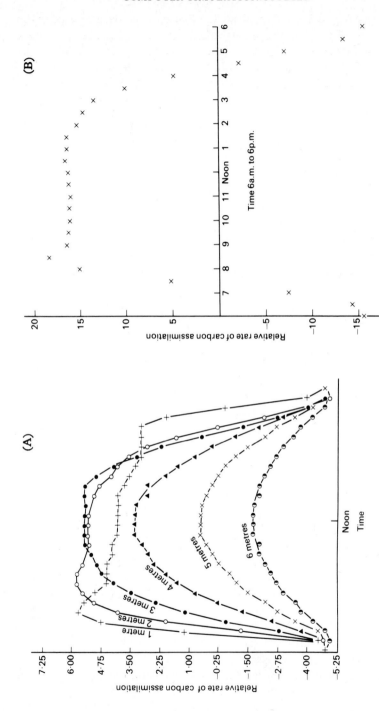

Fig. 12.20. (A) Simulated rate curves at various depths during a standard light day with the hysteresis effect incorporated into the model. (B) Observed rate curve at a simulated depth of one metre.

light energy available for photosynthesis at each of six lake depths. An initial value of RMAX, the largest value of energy input during the time period simulated; and RLAST, the last value of RZ (TIME); are initially set at 0·0 (Fig. 12.21). This allows the model to run on the upward part of the loop (Fig. 12.19) utilizing EQUATION 1, at the same time keeping a running check of RLAST. Subsequently if the incoming

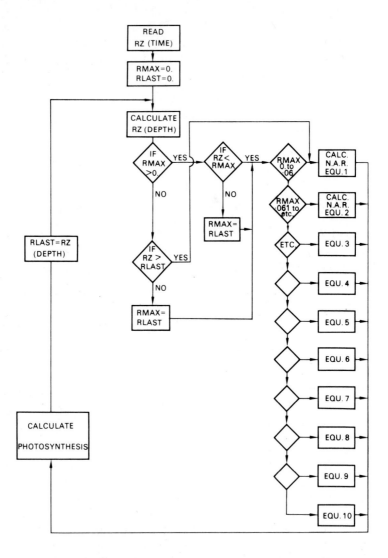

Fig. 12.21. Flow diagram of the model segment simulating photosynthesis in *Cosmarium*, with nine hysteresis return pathways.

light energy is reduced, RZ (TIME) is less than RLAST, RMAX is incremented, and the subsequent pathway through the model is via nine further equations dependent on the current value of RMAX (Fig. 12.21). These differing response pathways were determined experimentally.

This case study illustrates very well how a simple assumption can be misleading and re-emphasizes the importance of thorough testing of each assumption before it is built into the developing model. It also demonstrates how the hysteresis effect, which is a continuously varying situation, can be treated at a simple level by using nine equations to describe the 'average' response in nine 'sections' within the continuous range. Further breakdown into a greater number of sections ignores the inherent error level in the experimental data which was used initially to develop the primary hysteresis equations, and equally the error in the remaining sections of the model. Thus it is pointless to operate one section of the model to two decimal places accuracy when another part of the model is operating at the level of the 'nearest whole number'.

Case Study 5—Feedback From the Model

A simulation model attempts to answer a number of specific questions to which an experimental answer is either difficult or impossible—there is rarely any merit in constructing a model purely for its own sake. In addition to the specific answer sought, there always emerges a series of 'bonuses' in the form of feedback from the model. This feedback can be either negative or positive. Thus a simple combination of parameters may

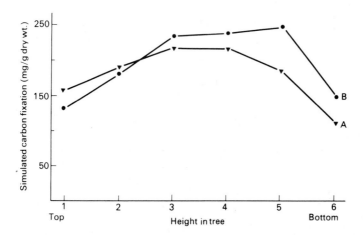

Fig. 12.22. (A) Simulated carbon fixation at six heights in a model tree using meteorological data for Plymouth. (B) Simulated carbon fixation under an increased evaporation gradient. (From Kershaw and Harris 1969; courtesy of Penn. State Univ. Press.)

describe a situation in a segment of the system where previously a more complex situation was thought to exist. Conversely, a component of the system containing the likely controlling equations *still* does not fit the experimental data, forcing a re-evaluation of this particular component and usually the uncovering of a totally unexpected relationship in the experimental situation.

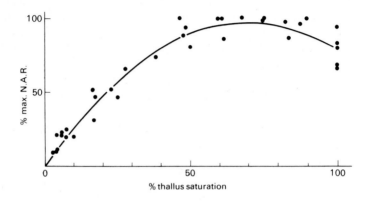

Fig. 12.23. The interaction between thallus water content and N. A. R. from five lichen replicates of *Parmelia caperata*. (From Kershaw and Harris 1969; courtesy of Penn. State Univ. Press.)

The data of Harris (1971*a* and *b*) formulated as a model (Harris 1972) demonstrates very well a bonus feedback from the provisional model Kershaw and Harris (1969). The model was constructed to test the hypothesis that light and water control the vertical distribution of the lichen *Parmelia caperata* on oak in S.W. England. The model showed that the simulated lichen distribution was in fact considerably altered by a change in the evaporation gradient of the model (Fig. 12.22) and gave reasonable agreement with the observed zonation. The control of growth of the lichen is simply related to the rate of carbon fixation being dependent on the degree of saturation of the thallus (Fig. 12.23), and hence the evaporation gradient. Using the model with different meteorological data from Boscombe Down in S.E. England showed a very different growth increment (see Fig. 12.24). The rainfall in S.E. England is considerably less than Devon in the West so, in part, this reduced carbon fixation is to be expected. However, *Parmelia caperata* occurs frequently in rural areas of Kent and Surrey, but it only has high cover values at the bases of very old trees in open woodland or along hedge rows. One is thus forced to the conclusion that as a general model it is incomplete and cannot be extrapolated outside the original defined system.

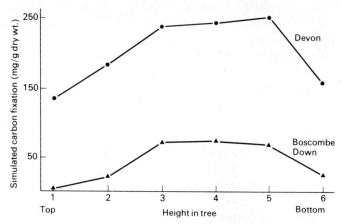

Fig. 12.24. Simulated carbon fixation using meteorological data for Plymouth compared with Boscombe Down. (From Kershaw and Harris 1969; courtesy of Penn. State Univ. Press.)

However, the model basically consists of the inter-relations between net radiation, and water availability at different heights, with the physiology of the lichen. Extrapolating from the Devon model would certainly not alter the physical laws governing net radiation, and water availability, but the physiology of the lichen could possibly be adapted to the less

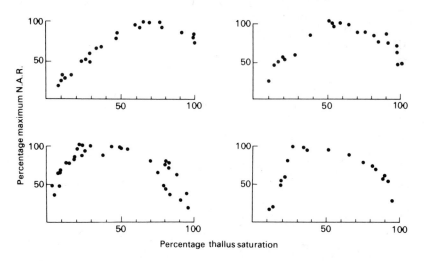

Fig. 12.25. Interaction between net assimilation rate and thallus water content for *Parmelia caperata* collected from (a) Devon, (b) Wiltshire, (c) Surrey, (d) Norfolk. (From Kershaw and Harris 1969; courtesy of Penn. State Univ. Press.)

mesic conditions in the southeast. This possibility was subsequently examined experimentally by monitoring net assimilation rates at different levels of thallus saturation, in a series of replicates collected from different parts of southern England. The results for Wiltshire, Surrey and Norfolk compared with those for Devon show striking differences in the position of their net assimilation optima (Fig. 12.25). A comparison with the rainfall distribution in southern England (Fig. 12.26) shows an obvious correlation of rainfall with the level of thallus saturation at which the net assimilation rate reaches an optimum. *Parmelia caperata* as a species either consists of a number of physiological strains each of which fills a suitable ecological niche, or is represented as a single strain which becomes fully adapted physiologically to less mesic habitats than its norm.

This finding was a totally unexpected feedback from the model and subsequent work has confirmed the physiological variability of *P. caperata* and shown it to be a widespread phenomena in lichens.

Over 50 cm rain

38 cm isohyet

Fig. 12.26. Rainfall distribution in southern England and optimal water contents for maximum net assimilation for thallus of *Parmelia caperata*. (From Kershaw and Harris 1969; courtesy of Penn. State Univ. Press.)

DISCUSSION

It is clear that simulation studies will advance rapidly in both complexity and extent. It is equally clear that they are an extremely powerful research tool—but sight of the objectives must not be lost during admiration of the elegance of the approach. There is often a tendency to concentrate purely on techniques at the expense of ecological information. This is true of the development of ordination methods as it is potentially true of the 1973 situation in modelling. Thus systems analysis has never been exactly defined Dale (1969) and its range of use extends from highly mathematical techniques to a simple but logical dissection of the sequential relationships and interactions of a series of parameters in a system. It may be very elegant to use Laplace transforms instead of a series of differential equations involved in a systems model, but if the mathematical background of the student is deficient, it is just as ecologically informative to build the model from very simple components with their inter-relationships arranged and controlled by standard computer iterative methods. The biologist with very little mathematics should not be overawed by the technical jargon. If differential equations are not understood then use simple *difference* equations. It is the validity of the ecological answer from the model which is important; the actual simulation model is merely a research tool, not an end in itself.

Appendix 1*
Distribution of t

Probability

n	·05	·02	·01	·001
1	12·706	31·821	63·657	636·619
2	4·303	6·965	9·925	31·598
3	3·182	4·541	5·841	12·941
4	2·776	3·747	4·604	8·610
5	2·571	3·365	4·032	6·859
6	2·447	3·143	3·707	5·959
7	2·365	2·998	3·499	5·405
8	2·306	2·896	3·355	5·041
9	2·262	2·821	3·250	4·781
10	2·228	2·764	3·169	4·587
11	2·201	2·718	3·106	4·437
12	2·179	2·681	3·055	4·318
13	2·160	2·650	3·012	4·221
14	2·145	2·624	2·977	4·140
15	2·131	2·602	2·947	4·073
16	2·120	2·583	2·921	4·015
17	2·110	2·567	2·898	3·965
18	2·101	2·552	2·878	3·922
19	2·093	2·539	2·861	3·883
20	2·086	2·528	2·845	3·850
21	2·080	2·518	2·831	3·819
22	2·074	2·508	2·819	3·792
23	2·069	2·500	2·807	3·767
24	2·064	2·492	2·797	3·745
25	2·060	2·485	2·787	3·725
26	2·056	2·479	2·779	3·707
27	2·052	2·473	2·771	3·690
28	2·048	2·467	2·763	3·674
29	2·045	2·462	2·756	3·659
30	2·042	2·457	2·750	3·646
40	2·021	2·423	2·704	3·551
60	2·000	2·390	2·660	3·460
120	1·980	2·358	2·617	3·373
∞	1·960	2·326	2·576	3·291

From Fisher and Yates; courtesy of Longman Group Ltd., London.

* The tables on pages 274–285 are taken from Fisher and Yates *Statistical Tables for Biological, Agricultural and Medical Research*, published by Longman Group Ltd., London (previously published by Oliver and Boyd, Edinburgh), and by permission of the authors and publishers.

Appendix 2
Distribution of χ^2

n	·05	·02	·01	·001
1	3·841	5·412	6·635	10·827
2	5·991	7·824	9·210	13·815
3	7·815	9·837	11·345	16·268
4	9·488	11·668	13·277	18·465
5	11·070	13·388	15·086	20·517
6	12·592	15·033	16·812	22·457
7	14·067	16·622	18·475	24·322
8	15·507	18·168	20·090	26·125
9	16·919	19·679	21·666	27·877
10	18·307	21·161	23·209	29·588
11	19·675	22·618	24·725	31·264
12	21·026	24·054	26·217	32·909
13	22·362	25·472	27·688	34·528
14	23·685	26·873	29·141	36·123
15	24·996	28·259	30·578	37·697
16	26·296	29·633	32·000	39·252
17	27·587	30·995	33·409	40·790
18	28·869	32·346	34·805	42·312
19	30·144	33·687	36·191	43·820
20	31·410	35·020	37·566	45·315
21	32·671	36·343	38·932	46·797
22	33·924	37·659	40·289	48·268
23	35·172	38·968	41·638	49·728
24	36·415	40·270	42·980	51·179
25	37·652	41·566	44·314	52·620
26	38·885	42·856	45·642	54·052
27	40·113	44·140	46·963	55·476
28	41·337	45·419	48·278	56·893
29	42·557	46·693	49·588	58·302
30	43·773	47·962	50·892	59·703

From Fisher and Yates; courtesy of Longman Group Ltd., London.

Appendix 3
Variance Ratio

5 per cent points of e^{2z}

n_2 \ n_1	1	2	3	4	5	6	8	12	24	∞
1	161·4	199·5	215·7	224·6	230·2	234·0	238·9	243·9	249·0	254·3
2	18·51	19·00	19·16	19·25	19·30	19·33	19·37	19·41	19·45	19·50
3	10·13	9·55	9·28	9·12	9·01	8·94	8·84	8·74	8·64	8·53
4	7·71	6·94	6·59	6·39	6·26	6·16	6·04	5·91	5·77	5·63
5	6·61	5·79	5·41	5·19	5·05	4·95	4·82	4·68	4·53	4·36
6	5·99	5·14	4·76	4·53	4·39	4·28	4·15	4·00	3·84	3·67
7	5·59	4·74	4·35	4·12	3·97	3·87	3·73	3·57	3·41	3·23
8	5·32	4·46	4·07	3·84	3·69	3·58	3·44	3·28	3·12	2·93
9	5·12	4·26	3·86	3·63	3·48	3·37	3·23	3·07	2·90	2·71
10	4·96	4·10	3·71	3·48	3·33	3·22	3·07	2·91	2·74	2·54
11	4·84	3·98	3·59	3·36	3·20	3·09	2·95	2·79	2·61	2·40
12	4·75	3·88	3·49	3·26	3·11	3·00	2·85	2·69	2·50	2·30
13	4·67	3·80	3·41	3·18	3·02	2·92	2·77	2·60	2·42	2·21
14	4·60	3·74	3·34	3·11	2·96	2·85	2·70	2·53	2·35	2·13
15	4·54	3·68	3·29	3·06	2·90	2·79	2·64	2·48	2·29	2·07
16	4·49	3·63	3·24	3·01	2·85	2·74	2·59	2·42	2·24	2·01
17	4·45	3·59	3·20	2·96	2·81	2·70	2·55	2·38	2·19	1·96
18	4·41	3·55	3·16	2·93	2·77	2·66	2·51	2·34	2·15	1·92
19	4·38	3·52	3·13	2·90	2·74	2·63	2·48	2·31	2·11	1·88
20	4·35	3·49	3·10	2·87	2·71	2·60	2·45	2·28	2·08	1·84
21	4·32	3·47	3·07	2·84	2·68	2·57	2·42	2·25	2·05	1·81
22	4·30	3·44	3·05	2·82	2·66	2·55	2·40	2·23	2·03	1·78
23	4·28	3·42	3·03	2·80	2·64	2·53	2·38	2·20	2·00	1·76
24	4·26	3·40	3·01	2·78	2·62	2·51	2·36	2·18	1·98	1·73
25	4·24	3·38	2·99	2·76	2·60	2·49	2·34	2·26	1·96	1·71
26	4·22	3·37	2·98	2·74	2·59	2·47	2·32	2·15	1·95	1·69
27	4·21	3·35	2·96	2·73	2·57	2·46	2·30	2·13	1·93	1·67
28	4·20	3·34	2·95	2·71	2·56	2·44	2·29	2·12	1·91	1·65
29	4·18	3·33	2·93	2·70	2·54	2·43	2·28	2·10	1·90	1·64
30	4·17	3·32	2·92	2·69	2·53	2·42	2·27	2·09	1·89	1·62
40	4·08	3·23	2·84	2·61	2·45	2·34	2·18	2·00	1·79	1·51
60	4·00	3·15	2·76	2·52	2·37	2·25	2·10	1·92	1·70	1·39
120	3·92	3·07	2·68	2·45	2·29	2·17	2·02	1·83	1·61	1·25
∞	3·84	2·99	2·60	2·37	2·21	2·09	1·94	1·75	1·52	1·00

From Fisher and Yates; courtesy of Longman Group Ltd., London.

Variance Ratio—*continued*

1 per cent points of e^{2z}

n_2 \ n_1	1	2	3	4	5	6	8	12	24	∞
1	4052	4999	5403	5625	5764	5859	5981	6106	6234	6366
2	98·49	99·00	99·17	99·25	99·30	99·33	99·36	99·42	99·46	99·50
3	34·12	30·81	29·46	28·71	28·24	27·91	27·49	27·05	26·60	26·12
4	21·20	18·00	16·69	15·98	15·52	15·21	14·80	14·37	13·93	13·46
5	16·26	13·27	12·06	11·39	10·97	10·67	10·29	9·89	9·47	9·02
6	13·74	10·92	9·78	9·15	8·75	8·47	8·10	7·72	7·31	6·88
7	12·25	9·55	8·45	7·85	7·46	7·19	6·84	6·47	6·07	5·65
8	11·26	8·65	7·59	7·01	6·63	6·37	6·03	5·67	5·28	4·86
9	10·56	8·02	6·99	6·42	6·06	5·80	5·47	5·11	4·73	4·31
10	10·04	7·56	6·55	5·99	5·64	5·39	5·06	4·71	4·33	3·91
11	9·65	7·20	6·22	5·67	5·32	5·07	4·74	4·40	4·02	3·60
12	9·33	6·93	5·95	5·41	5·06	4·82	4·50	4·16	3·78	3·36
13	9·07	6·70	5·74	5·20	4·86	4·62	4·30	3·96	3·59	3·16
14	8·86	6·51	5·56	5·03	4·69	4·46	4·14	3·80	3·43	3·00
15	8·68	6·36	5·42	4·89	4·56	4·32	4·00	3·67	3·29	2·87
16	8·53	6·23	5·29	4·77	4·44	4·20	3·89	3·55	3·18	2·75
17	8·40	6·11	5·18	4·67	4·34	4·10	3·79	3·45	3·08	2·65
18	8·28	6·01	5·09	4·58	4·25	4·01	3·71	3·37	3·00	2·57
19	8·18	5·93	5·01	4·50	4·17	3·94	3·63	3·30	2·92	2·49
20	8·10	5·85	4·94	4·43	4·10	3·87	3·56	3·23	2·86	2·42
21	8·02	5·78	4·87	4·37	4·04	3·81	3·51	3·17	2·80	2·36
22	7·94	5·72	4·82	4·31	3·99	3·76	3·45	3·12	2·75	2·31
23	7·88	5·66	4·76	4·26	3·94	3·71	3·41	3·07	2·70	2·26
24	7·82	5·61	4·72	4·22	3·90	3·67	3·36	3·03	2·66	2·21
25	7·77	5·57	4·68	4·18	3·86	3·63	3·32	2·99	2·62	2·17
26	7·72	5·53	4·64	4·14	3·82	3·59	3·29	2·96	2·58	2·13
27	7·68	5·49	4·60	4·11	3·78	3·56	3·26	2·93	2·55	2·10
28	7·64	5·45	4·57	4·07	3·75	3·53	3·23	2·90	2·52	2·06
29	7·60	5·42	4·54	4·04	3·73	3·50	3·20	2·87	2·49	2·03
30	7·56	5·39	4·51	4·02	3·70	3·47	3·17	2·84	2·47	2·01
40	7·31	5·18	4·31	3·83	3·51	3·29	2·99	2·66	2·29	1·80
60	7·08	4·98	4·13	3·65	3·34	3·12	2·82	2·50	2·12	1·60
120	6·85	4·79	3·95	3·48	3·17	2·96	2·66	2·34	1·95	1·38
∞	6·64	4·60	3·78	3·32	3·02	2·80	2·51	2·18	1·79	1·00

From Fisher and Yates; courtesy of Longman Group Ltd., London.

Variance Ratio—*continued*

0·1 per cent points of e^{2z}

n_2 \ n_1	1	2	3	4	5	6	8	12	24	∞
1	405284	500000	540379	562500	576405	585937	598144	610667	623497	636619
2	998·5	999·0	999·2	999·2	999·3	999·3	999·4	999·4	999·5	999·5
3	167·5	148·5	141·1	137·1	134·6	132·8	130·6	128·3	125·9	123·5
4	74·14	61·25	56·18	53·44	51·71	50·53	49·00	47·41	45·77	44·05
5	47·04	36·61	33·20	31·09	29·75	28·84	27·64	26·42	25·14	23·78
6	35·51	27·00	23·70	21·90	20·81	20·03	19·03	17·99	16·89	15·75
7	29·22	21·69	18·77	17·19	16·21	15·52	14·63	13·71	12·73	11·69
8	25·42	18·49	15·83	14·39	13·49	12·86	12·04	11·19	10·30	9·34
9	22·86	16·39	13·90	12·56	11·71	11·13	10·37	9·57	8·72	7·81
10	21·04	14·91	12·55	11·28	10·48	9·92	9·20	8·45	7·64	6·76
11	19·69	13·81	11·56	10·35	9·58	9·05	8·35	7·63	6·85	6·00
12	18·64	12·97	10·80	9·63	8·89	8·38	7·71	7·00	6·25	5·42
13	17·81	12·31	10·21	9·07	8·35	7·86	7·21	6·52	5·78	4·97
14	17·14	11·78	9·73	8·62	7·92	7·43	6·80	6·13	5·41	4·60
15	16·59	11·34	9·34	8·25	7·57	7·09	6·47	5·81	5·10	4·31
16	16·12	10·97	9·00	7·94	7·27	6·81	6·19	5·55	4·85	4·06
17	15·72	10·66	8·73	7·68	7·02	6·56	5·96	5·32	4·63	3·85
18	15·38	10·39	8·49	7·46	6·81	6·35	5·76	5·13	4·45	3·67
19	15·08	10·16	8·28	7·26	6·61	6·18	5·59	4·97	4·29	3·52
20	14·82	9·95	8·10	7·10	6·46	6·02	5·44	4·82	4·15	3·38
21	14·59	9·77	7·94	6·95	6·32	5·88	5·31	4·70	4·03	3·26
22	14·38	9·61	7·80	6·81	6·19	5·76	5·19	4·58	3·92	3·15
23	14·19	9·47	7·67	6·69	6·08	5·65	5·09	4·48	3·82	3·05
24	14·03	9·34	7·55	6·59	5·98	5·55	4·99	4·39	3·74	2·97
25	13·88	9·22	7·45	6·49	5·88	5·46	4·91	4·31	3·66	2·89
26	13·74	9·12	7·36	6·41	5·80	5·38	4·83	4·24	3·59	2·82
27	13·61	9·02	7·27	6·33	5·73	5·31	4·76	4·17	3·52	2·75
28	13·50	8·93	7·19	6·25	5·66	5·24	4·69	4·11	3·46	2·70
29	13·39	8·85	7·12	6·19	5·59	5·18	4·64	4·05	3·41	2·64
30	13·29	8·77	7·05	6·12	5·53	5·12	4·58	4·00	3·36	2·59
40	12·61	8·25	6·60	5·70	5·13	4·73	4·21	3·64	3·01	2·23
60	11·97	7·76	6·17	5·31	4·76	4·37	3·87	3·31	2·69	1·90
120	11·38	7·31	5·79	4·95	4·42	4·04	3·55	3·02	2·40	1·56
∞	10·83	6·91	5·42	4·62	4·10	3·74	3·27	2·74	2·13	1·00

From Fisher and Yates; courtesy of Longman Group Ltd., London.

Appendix 4

Values of the correlation coefficient for different levels of significance

n	·1	·05	·02	·01	·001
1	·98769	·99692	·999507	·999877	·9999988
2	·90000	·95000	·98000	·990000	·99900
3	·8054	·8783	·93433	·95873	·99116
4	·7293	·8114	·8822	·91720	·97406
5	·6694	·7545	·8329	·8745	·95074
6	·6215	·7067	·7887	·8343	·92493
7	·5822	·6664	·7498	·7977	·8982
8	·5494	·6319	·7155	·7646	·8721
9	·5214	·6021	·6851	·7348	·8471
10	·4973	·5760	·6581	·7079	·8233
11	·4762	·5529	·6339	·6835	·8010
12	·4575	·5324	·6120	·6614	·7800
13	·4409	·5139	·5923	·6411	·7603
14	·4259	·4973	·5742	·6226	·7420
15	·4124	·4821	·5577	·6055	·7246
16	·4000	·4683	·5425	·5897	·7084
17	·3887	·4555	·5285	·5751	·6932
18	·3783	·4438	·5155	·5614	·6787
19	·3687	·4329	·5034	·5487	·6652
20	·3598	·4227	·4921	·5368	·6524
25	·3233	·3809	·4451	·4869	·5974
30	·2960	·3494	·4093	·4487	·5541
35	·2746	·3246	·3810	·4182	·5189
40	·2573	·3044	·3578	·3932	·4896
45	·2428	·2875	·3384	·3721	·4648
50	·2306	·2732	·3218	·3541	·4433
60	·2108	·2500	·2948	·3248	·4078
70	·1954	·2319	·2727	·3017	·3799
80	·1829	·2172	·2565	·2830	·3568
90	·1726	·2050	·2427	·2673	·3375
100	·1638	·1946	·2301	·2540	·3211

From Fisher and Yates; courtesy of Longman Group Ltd., London.

Appendix 5

Random numbers (I)

03 47 43 73 86	36 96 47 36 61	46 98 63 71 62	33 26 16 80 45	60 11 14 10 95
97 74 24 67 62	42 81 14 57 20	42 53 32 37 32	27 07 36 07 51	24 51 79 89 73
16 76 62 27 66	56 50 26 71 07	32 90 79 78 53	13 55 38 58 59	88 97 54 14 10
12 56 85 99 26	96 96 68 27 31	05 03 72 93 15	57 12 10 14 21	88 26 49 81 76
55 59 56 35 64	38 54 82 46 22	31 62 43 09 90	06 18 44 32 53	23 83 01 30 30
16 22 77 94 39	49 54 43 54 82	17 37 93 23 78	87 35 20 96 43	84 26 34 91 64
84 42 17 53 31	57 24 55 06 88	77 04 74 47 67	21 76 33 50 25	83 92 12 06 76
63 01 63 78 59	16 95 55 67 19	98 10 50 71 75	12 86 73 58 07	44 39 52 38 79
33 21 12 34 29	78 64 56 07 82	52 42 07 44 38	15 51 00 13 42	99 66 02 79 54
57 60 86 32 44	09 47 27 96 54	49 17 46 09 62	90 52 84 77 27	08 02 73 43 28
18 18 07 92 46	44 17 16 58 09	79 83 86 19 62	06 76 50 03 10	55 23 64 05 05
26 62 38 97 75	84 16 07 44 99	83 11 46 32 24	20 14 85 88 45	10 93 72 88 71
23 42 40 64 74	82 97 77 77 81	07 45 32 14 08	32 98 94 07 72	93 85 79 10 75
52 36 28 19 95	50 92 26 11 97	00 56 76 31 38	80 22 02 53 53	86 60 42 04 53
34 85 94 35 12	83 39 50 08 30	42 34 07 96 88	54 42 06 87 98	35 85 29 48 39
70 29 17 12 13	40 33 20 38 26	13 89 51 03 74	17 76 37 13 04	07 74 21 19 30
56 62 18 37 35	96 83 50 87 75	97 12 25 93 47	70 33 24 03 54	97 77 46 44 80
99 49 57 22 77	88 42 95 45 72	16 64 36 16 00	04 43 18 66 79	94 77 24 21 90
16 08 15 04 72	33 27 14 34 09	45 59 34 68 49	12 72 07 34 45	99 27 72 95 14
31 16 93 32 43	50 27 89 87 19	20 15 37 00 49	52 85 66 60 44	38 68 88 11 80
68 34 30 13 70	55 74 30 77 40	44 22 78 84 26	04 33 46 09 52	68 07 97 06 57
74 57 25 65 76	59 29 97 68 60	71 91 38 67 54	13 58 18 24 76	15 54 55 95 52
27 42 37 86 53	48 55 90 65 72	96 57 69 36 10	96 46 92 42 45	97 60 49 04 91
00 39 68 29 61	66 37 32 20 30	77 84 57 03 29	10 45 65 04 26	11 04 96 67 24
29 94 98 94 24	68 49 69 10 82	53 75 91 93 30	34 25 20 57 27	40 48 73 51 92
16 90 82 66 59	83 62 64 11 12	67 19 00 71 74	60 47 21 29 68	02 02 37 03 31
11 27 94 75 06	06 09 19 74 66	02 94 37 34 02	76 70 90 30 86	38 45 94 30 38
35 24 10 16 20	33 32 51 26 38	79 78 45 04 91	16 92 53 56 16	02 75 50 95 98
38 23 16 86 38	42 38 97 01 50	87 75 66 81 41	40 01 74 91 62	48 51 84 08 32
31 96 25 91 47	96 44 33 49 13	34 86 82 53 91	00 52 43 48 85	27 55 26 89 62
66 67 40 67 14	64 05 71 95 86	11 05 65 09 68	76 83 20 37 90	57 16 00 11 66
14 90 84 45 11	75 73 88 05 90	52 27 41 14 86	22 98 12 22 08	07 52 74 95 80
68 05 51 18 00	33 96 02 75 19	07 60 62 93 55	59 33 82 43 90	49 37 38 44 59
20 46 78 73 90	97 51 40 14 02	04 02 33 31 08	39 54 16 49 36	47 95 93 13 30
64 19 58 97 79	15 06 15 93 20	01 90 10 75 06	40 78 78 89 62	02 67 74 17 33
05 26 93 70 60	22 35 85 15 13	92 03 51 59 77	59 56 78 06 83	52 91 05 70 74
07 97 10 88 23	09 98 42 99 64	61 71 62 99 15	06 51 29 16 93	58 05 77 09 51
68 71 86 85 85	54 87 66 47 54	73 32 08 11 12	44 95 92 63 16	29 56 24 29 48
26 99 61 65 53	58 37 78 80 70	42 10 50 67 42	32 17 55 85 74	94 44 67 16 94
14 65 52 68 75	87 59 36 22 41	26 78 63 06 55	13 08 27 01 50	15 29 39 39 43
17 53 77 58 71	71 41 61 50 72	12 41 94 96 26	44 95 27 36 99	02 96 74 30 83
90 26 59 21 19	23 52 23 33 12	96 93 02 18 39	07 02 18 36 07	25 99 32 70 23
41 23 52 55 99	31 04 49 69 96	10 47 48 45 88	13 41 43 89 20	97 17 14 49 17
60 20 50 81 69	31 99 73 68 68	35 81 33 03 76	24 30 12 48 60	18 99 10 72 34
91 25 38 05 90	94 58 28 41 36	45 37 59 03 09	90 35 57 29 12	82 62 54 65 60
34 50 57 74 37	98 80 33 00 91	09 77 93 19 82	74 94 80 04 04	45 07 31 66 49
85 22 04 39 43	73 81 53 94 79	33 62 46 86 28	08 31 54 46 31	53 94 13 38 47
09 79 13 77 48	73 82 97 22 21	05 03 27 24 83	72 89 44 05 60	35 80 39 94 88
88 75 80 18 14	22 95 75 42 49	39 32 82 22 49	02 48 07 70 37	16 04 61 67 87
90 96 23 70 00	39 00 03 06 90	55 85 78 38 36	94 37 30 69 32	90 89 00 76 33

From Fisher and Yates; courtesy of Longman Group Ltd., London.

Random numbers (II)

53 74 23 99 67	61 32 28 69 84	94 62 67 86 24	98 33 41 19 95	47 53 53 38 09
63 38 06 86 54	99 00 65 26 94	02 82 90 23 07	79 62 67 80 60	75 91 12 81 19
35 30 58 21 46	06 72 17 10 94	25 21 31 75 96	49 28 24 00 49	55 65 79 78 07
63 43 36 82 69	65 51 18 37 88	61 38 44 12 45	32 92 85 88 65	54 34 81 85 35
98 25 37 55 26	01 91 82 81 46	74 71 12 94 97	24 02 71 37 07	03 92 18 66 75
02 63 21 17 69	71 50 80 89 56	38 15 70 11 48	43 40 45 86 98	00 83 26 91 03
64 55 22 21 82	48 22 28 06 00	61 54 13 43 91	82 78 12 23 29	06 66 24 12 27
85 07 26 13 89	01 10 07 82 04	59 63 69 36 03	69 11 15 83 80	13 29 54 19 28
58 54 16 24 15	51 54 44 82 00	62 61 65 04 69	38 18 65 18 97	85 72 13 49 21
34 85 27 84 87	61 48 64 56 26	90 18 48 13 26	37 70 15 42 57	65 65 80 39 07
03 92 18 27 46	57 99 16 96 56	30 33 72 85 22	84 64 38 56 98	99 01 30 98 64
62 95 30 27 59	37 75 41 66 48	86 97 80 61 45	23 53 04 01 63	45 76 08 64 27
08 45 93 15 22	60 21 75 46 91	98 77 27 85 42	28 88 61 08 84	69 62 03 42 73
07 08 55 18 40	45 44 75 13 90	24 94 96 61 02	57 55 66 83 15	73 42 37 11 61
01 85 89 95 66	51 10 19 34 88	15 84 97 19 75	12 76 39 43 78	64 63 91 08 25
72 84 71 14 35	19 11 58 49 26	50 11 17 17 76	86 31 57 20 18	95 60 78 46 75
88 78 28 16 84	13 52 53 94 53	75 45 69 30 96	73 89 65 70 31	99 17 43 48 76
45 17 75 65 57	28 40 19 72 12	25 12 74 75 67	60 40 60 81 19	24 62 01 61 16
96 76 28 12 54	22 01 11 94 25	71 96 16 16 88	68 64 36 74 45	19 59 50 88 92
43 31 67 72 30	24 02 94 08 63	38 32 36 66 02	69 36 38 25 39	48 03 45 15 22
50 44 66 44 21	66 06 58 05 62	68 15 54 35 02	42 35 48 96 32	14 52 41 52 48
22 66 22 15 86	26 63 75 41 99	58 42 36 72 24	58 37 52 18 51	03 37 18 39 11
96 24 40 14 51	23 22 30 88 57	95 67 47 29 83	94 69 40 06 07	18 16 36 78 86
31 73 91 61 19	60 20 72 93 48	98 57 07 23 69	65 95 39 69 58	56 80 30 19 44
78 60 73 99 84	43 89 94 36 45	56 69 47 07 41	90 22 91 07 12	78 35 34 08 72
84 37 90 61 56	70 10 23 98 05	85 11 34 76 60	76 48 45 34 60	01 64 18 39 96
36 67 10 08 23	98 93 35 08 86	99 29 76 29 81	33 34 91 58 93	63 14 52 32 52
07 28 59 07 48	89 64 58 89 75	83 85 62 27 89	30 14 78 56 27	86 63 59 80 02
10 15 83 87 60	79 24 31 66 56	21 48 24 06 93	91 98 94 05 49	01 47 59 38 00
55 19 68 97 65	03 73 52 16 56	00 53 55 90 27	33 42 29 38 87	22 18 88 83 34
53 81 29 13 39	35 01 20 71 34	62 33 74 82 14	53 73 19 09 03	56 54 29 56 93
51 86 32 68 92	33 98 74 66 99	40 14 71 94 58	45 94 19 38 81	14 44 99 81 07
35 91 70 29 13	80 03 54 07 27	96 94 78 32 66	50 95 52 74 33	13 80 55 62 54
37 71 67 95 13	20 02 44 95 94	64 85 04 05 72	01 32 90 76 14	53 89 74 60 41
93 66 13 83 27	92 79 64 64 72	28 54 96 53 84	48 14 52 98 94	56 07 93 89 30
02 96 08 45 65	13 05 00 41 84	93 07 54 72 59	21 45 57 09 77	19 48 56 27 44
49 83 43 48 35	82 88 33 69 96	72 36 04 19 76	47 45 15 18 60	82 11 08 95 97
84 60 71 62 46	40 80 81 30 37	34 39 23 05 38	25 15 35 71 30	88 12 57 21 77
18 17 30 88 71	44 91 14 88 47	89 23 30 63 15	56 34 20 47 89	99 82 93 24 98
79 69 10 61 78	71 32 76 95 62	87 00 22 58 40	92 54 01 75 25	43 11 71 99 31
75 93 36 57 83	56 20 14 82 11	74 21 97 90 65	96 42 68 63 86	74 54 13 26 94
38 30 92 29 03	06 28 81 39 38	62 25 06 84 63	61 29 08 93 67	04 32 92 08 09
51 29 50 10 34	31 57 75 95 80	51 97 02 74 77	76 15 48 49 44	18 55 63 77 09
21 31 38 86 24	37 79 81 53 74	73 24 16 10 33	52 83 90 94 76	70 47 14 54 36
29 01 23 87 88	58 02 39 37 67	42 10 14 20 92	16 55 23 42 45	54 96 09 11 06
95 33 95 22 00	18 74 72 00 18	38 79 58 69 32	81 76 80 26 92	82 80 84 25 39
90 84 60 79 80	24 36 59 87 38	82 07 53 89 35	96 35 23 79 18	05 98 90 07 35
46 40 62 98 82	54 97 20 56 95	15 74 80 08 32	16 46 70 50 80	67 72 16 42 79
20 31 89 03 43	38 46 82 68 72	32 14 82 99 70	80 60 47 18 97	63 49 30 21 30
71 59 73 05 50	08 22 23 71 77	91 01 93 20 49	82 96 59 26 94	66 39 67 98 60

From Fisher and Yates; courtesy of Longman Group Ltd., London.

Random numbers (III)

```
22 17 68 65 84   68 95 23 92 35   87 02 22 57 51   61 09 43 95 06   58 24 82 03 47
19 36 27 59 46   13 79 93 37 55   39 77 32 77 09   85 52 05 30 62   47 83 51 62 74
16 77 23 02 77   09 61 87 25 21   28 06 24 25 93   16 71 13 59 78   23 05 47 47 25
78 43 76 71 61   20 44 90 32 64   97 67 63 99 61   46 38 03 93 22   69 81 21 99 21
03 28 28 26 08   73 37 32 04 05   69 30 16 09 05   88 69 58 28 99   35 07 44 73 47

93 22 53 64 39   07 10 63 76 35   87 03 04 79 88   08 13 13 85 51   55 34 57 72 69
78 76 58 54 74   92 38 70 96 92   52 06 79 79 45   82 63 18 27 44   69 66 92 19 09
23 68 35 26 00   99 53 93 61 28   52 70 05 48 34   56 65 05 61 86   90 92 10 70 80
15 39 25 70 99   93 86 52 77 65   15 33 59 05 28   22 87 26 07 47   86 96 98 29 06
58 71 96 30 24   18 46 23 34 27   85 13 99 24 44   49 18 09 79 49   74 16 32 23 02

57 35 27 33 72   24 53 63 94 09   41 10 76 47 91   44 04 95 49 66   39 60 04 59 81
48 50 86 54 48   22 06 34 72 52   82 21 15 65 20   33 29 94 71 11   15 91 29 12 03
61 96 48 95 03   07 16 39 33 66   98 56 10 56 79   77 21 30 27 12   90 49 22 23 62
36 93 89 41 26   29 70 83 63 51   99 74 20 52 36   87 09 41 15 09   98 60 16 03 03
18 87 00 42 31   57 90 12 02 07   23 47 37 17 31   54 08 01 88 63   39 41 88 92 10

88 56 53 27 59   33 35 72 67 47   77 34 55 45 70   08 18 27 38 90   16 95 86 70 75
09 72 95 84 29   49 41 31 06 70   42 38 06 45 18   64 84 73 31 65   52 53 37 97 15
12 96 88 17 31   65 19 69 02 83   60 75 86 90 68   24 64 19 35 51   56 61 87 39 12
85 94 57 24 16   92 09 84 38 76   22 00 27 69 85   29 81 94 78 70   21 94 47 90 12
38 64 43 59 98   98 77 87 68 07   91 51 67 62 44   40 98 05 93 78   23 32 65 41 18

53 44 09 42 72   00 41 86 79 79   68 47 22 00 20   35 55 31 51 51   00 83 63 22 55
40 76 66 26 84   57 99 99 90 37   36 63 32 08 58   37 40 13 68 97   87 64 81 07 83
02 17 79 18 05   12 59 52 57 02   22 07 90 47 03   28 14 11 30 79   20 69 22 40 98
95 17 82 06 53   31 51 10 96 46   92 06 88 07 77   56 11 50 81 69   40 23 72 51 39
35 76 22 42 92   96 11 83 44 80   34 68 35 48 77   33 42 40 90 60   73 96 53 97 86

26 29 13 56 41   85 47 04 66 08   34 72 57 59 13   82 43 80 46 15   38 26 61 70 04
77 80 20 75 82   72 82 32 99 90   63 95 73 76 63   89 73 44 99 05   48 67 26 43 18
46 40 66 44 52   91 36 74 43 53   30 82 13 54 00   78 45 63 98 35   55 03 36 67 68
37 56 08 18 09   77 53 84 46 47   31 91 18 95 58   24 16 74 11 53   44 10 13 85 57
61 65 61 68 66   37 27 47 39 19   84 83 70 07 48   53 21 40 60 71   95 06 79 88 54

93 43 69 64 07   34 18 04 52 35   56 27 09 24 86   61 85 53 83 45   19 90 70 99 00
21 96 60 12 99   11 20 99 45 18   48 13 93 55 34   18 37 79 49 90   65 97 38 20 46
95 20 47 97 97   27 37 83 28 71   00 06 41 41 74   45 89 09 39 84   51 67 11 52 49
97 86 21 78 73   10 65 81 92 59   58 76 17 14 97   04 76 62 16 17   17 95 70 45 80
69 92 06 34 13   59 71 74 17 32   27 55 10 24 19   23 71 82 13 74   63 52 52 01 41

04 31 17 21 56   33 73 99 19 87   26 72 39 27 67   53 77 57 68 93   60 61 97 22 61
61 06 98 03 91   87 14 77 43 06   43 00 65 98 50   45 60 33 01 07   98 99 46 50 47
85 93 85 86 88   72 87 08 62 40   16 06 10 89 20   23 21 34 74 97   76 38 03 29 63
21 74 32 47 45   73 96 07 94 52   09 65 90 77 47   25 76 16 19 33   53 05 70 53 30
15 69 53 82 80   79 96 23 53 10   65 39 07 16 29   45 33 02 43 70   02 87 40 41 45

02 89 08 04 49   20 21 14 68 86   87 63 93 95 17   11 29 01 95 80   35 14 97 35 33
87 18 15 89 79   85 43 01 72 73   08 61 74 51 69   89 74 39 82 15   94 51 33 41 67
98 83 71 94 22   59 97 50 99 52   08 52 85 08 40   87 80 61 65 31   91 51 80 32 44
10 08 58 21 66   72 68 49 29 31   89 85 84 46 06   59 73 19 85 23   65 09 29 75 63
47 90 56 10 08   88 02 84 27 83   42 29 72 23 19   66 56 45 65 79   20 71 53 20 25

22 85 61 68 90   49 64 92 85 44   16 40 12 89 88   50 14 49 81 06   01 82 77 45 12
67 80 43 79 33   12 83 11 41 16   25 58 19 68 70   77 02 54 00 52   53 43 37 15 26
27 62 50 96 72   79 44 61 40 15   14 53 40 65 39   27 31 58 50 28   11 39 03 34 25
33 78 80 87 15   38 30 06 38 21   14 47 47 07 26   54 96 87 53 32   40 36 40 96 76
13 13 92 66 99   47 24 49 57 74   32 25 43 62 17   10 97 11 69 84   99 63 22 32 98
```

Random numbers (IV)

10 27 53 96 23	71 50 54 36 23	54 31 04 82 98	04 14 12 15 09	26 78 25 47 47
28 41 50 61 88	64 85 27 20 18	83 36 36 05 56	39 71 65 09 62	94 76 62 11 89
34 21 42 57 02	59 19 18 97 48	80 30 03 30 98	05 24 67 70 07	84 97 50 87 46
61 81 77 23 23	82 82 11 54 08	53 28 70 58 96	44 07 39 55 43	42 34 43 39 28
61 15 18 13 54	16 86 20 26 88	90 74 80 55 09	14 53 90 51 17	52 01 63 01 59
91 76 21 64 64	44 91 13 32 97	75 31 62 66 54	84 80 32 75 77	56 08 25 70 29
00 97 79 08 06	37 30 28 59 85	53 56 68 53 40	01 74 39 59 73	30 19 99 85 48
36 46 18 34 94	75 20 80 27 77	78 91 69 16 00	08 43 18 73 68	67 69 61 34 25
88 98 99 60 50	65 95 79 42 94	93 62 40 89 96	43 56 47 71 66	46 76 29 67 02
04 37 59 87 21	05 02 03 24 17	47 97 81 56 51	92 34 86 01 82	55 51 33 12 91
63 62 06 34 41	94 21 78 55 09	72 76 45 16 94	29 95 81 83 83	79 88 01 97 30
78 47 23 53 90	34 41 92 45 71	09 23 70 70 07	12 38 92 79 43	14 85 11 47 23
87 68 62 15 43	53 14 36 59 25	54 47 33 70 15	59 24 48 40 35	50 03 42 99 36
47 60 92 10 77	88 59 53 11 52	66 25 69 07 04	48 68 64 71 06	61 65 70 22 12
56 88 87 59 41	65 28 04 67 53	95 79 88 37 31	50 41 06 94 76	81 83 17 16 33
02 57 45 86 67	73 43 07 34 48	44 26 87 93 29	77 09 61 67 84	06 69 44 77 75
31 54 14 13 17	48 62 11 90 60	68 12 93 64 28	46 24 79 16 76	14 60 25 51 01
28 50 16 43 36	28 97 85 58 99	67 22 52 76 23	24 70 36 54 54	59 28 61 71 96
63 29 62 66 50	02 63 45 52 38	67 63 47 54 75	83 24 78 43 20	92 63 13 47 48
45 65 58 26 51	76 96 59 38 72	86 57 45 71 46	44 67 76 14 55	44 88 01 62 12
39 65 36 63 70	77 45 85 50 51	74 13 39 35 22	30 53 36 02 95	49 34 88 73 61
73 71 98 16 04	29 18 94 51 23	76 51 94 84 86	79 93 96 38 63	08 58 25 58 94
72 20 56 20 11	72 65 71 08 86	79 57 95 13 91	97 48 72 66 48	09 71 17 24 89
75 17 26 99 76	89 37 20 70 01	77 31 61 95 46	26 97 05 73 51	53 33 18 72 87
37 48 60 82 29	81 30 15 39 14	48 38 75 93 29	06 87 37 78 48	45 56 00 84 47
68 08 02 80 72	83 71 46 30 49	89 17 95 88 29	02 39 56 03 46	97 74 06 56 17
14 23 98 61 67	70 52 85 01 50	01 84 02 78 43	10 62 98 19 41	18 83 99 47 99
49 08 96 21 44	25 27 99 41 28	07 41 08 34 66	19 42 74 39 91	41 96 53 78 72
78 37 06 08 43	63 61 62 42 29	39 68 95 10 96	09 24 23 00 62	56 12 80 73 16
37 21 34 17 68	68 96 83 23 56	32 84 60 15 31	44 73 67 34 77	91 15 79 74 58
14 29 09 34 04	87 83 07 55 07	76 58 30 83 64	87 29 25 58 84	86 50 60 00 25
58 43 28 06 36	49 52 83 51 14	47 56 91 29 34	05 87 31 06 95	12 45 57 09 09
10 43 67 29 70	80 62 80 03 42	10 80 21 38 84	90 56 35 03 09	43 12 74 49 14
44 38 88 39 54	86 97 37 44 22	00 95 01 31 76	17 16 29 56 63	38 78 94 49 81
90 69 59 19 51	85 39 52 85 13	07 28 37 07 61	11 16 36 27 03	78 86 72 04 95
41 47 10 25 62	97 05 31 03 61	20 26 36 31 62	68 69 86 95 44	84 95 48 46 45
91 94 14 63 19	75 89 11 47 11	31 56 34 19 09	79 57 92 36 59	14 93 87 81 40
80 06 54 18 66	09 18 94 06 19	98 40 07 17 81	22 45 44 84 11	24 62 20 42 31
67 72 77 63 48	84 08 31 55 58	24 33 45 77 58	80 45 67 93 82	75 70 16 08 24
59 40 24 13 27	79 26 88 86 30	01 31 60 10 39	53 58 47 70 93	85 81 56 39 38
05 90 35 89 95	01 61 16 96 94	50 78 13 69 36	37 68 53 37 31	71 26 35 03 71
44 43 80 69 98	46 68 05 14 82	90 78 50 05 62	77 79 13 57 44	59 60 10 39 66
61 81 31 96 82	00 57 25 60 59	46 72 60 18 77	55 66 12 62 11	08 99 55 64 57
42 88 07 10 05	24 98 65 63 21	47 21 61 88 32	27 80 30 21 60	10 92 35 36 12
77 94 30 05 39	28 10 99 00 27	12 73 73 99 12	49 99 57 94 82	96 88 57 17 91
78 83 19 76 16	94 11 68 84 26	23 54 20 86 85	23 86 66 99 07	36 37 34 92 09
87 76 59 61 81	43 63 64 61 61	65 76 36 95 90	18 48 27 45 68	27 23 65 30 72
91 43 05 96 47	55 78 99 95 24	37 55 85 78 78	01 48 41 19 10	35 19 54 07 73
84 97 77 72 73	09 62 06 65 72	87 12 49 03 60	41 15 20 76 27	50 47 02 29 16
87 41 60 76 83	44 88 96 07 80	83 05 83 38 96	73 70 66 81 90	30 56 10 48 59

From Fisher and Yates; courtesy of Longman Group Ltd., London.

Random numbers (V)

28 89 65 87 08	13 50 63 04 23	25 47 57 91 13	52 62 24 19 94	91 67 48 57 10
30 29 43 65 42	78 66 28 55 80	47 46 41 90 08	55 98 78 10 70	49 92 05 12 07
95 74 62 60 53	51 57 32 22 27	12 72 72 27 77	44 67 32 23 13	67 95 07 76 30
01 85 54 96 72	66 86 65 64 60	56 59 75 36 75	46 44 33 63 71	54 50 06 44 75
10 91 46 96 86	19 83 52 47 53	65 00 51 93 51	30 80 05 19 29	56 23 27 19 03
05 33 18 08 51	51 78 57 26 17	34 87 96 23 95	89 99 93 39 79	11 28 94 15 52
04 43 13 37 00	79 68 96 26 60	70 39 83 66 56	62 03 55 86 57	77 55 33 62 02
05 85 40 25 24	73 52 93 70 50	48 21 47 74 63	17 27 27 51 26	35 96 29 00 45
84 90 90 65 77	63 99 25 69 02	09 04 03 35 78	19 79 95 07 21	02 84 48 51 97
28 55 53 09 48	86 28 30 02 35	71 30 32 06 47	93 74 21 86 33	49 90 21 69 74
89 83 40 69 80	97 96 47 59 97	56 33 24 87 36	17 18 16 90 46	75 27 28 52 13
73 20 96 05 68	93 41 69 96 07	97 50 81 79 59	42 37 13 81 83	92 42 85 04 31
10 89 07 76 21	40 24 74 36 42	40 33 04 46 24	35 63 02 31 61	34 59 43 36 96
91 50 27 78 37	06 06 16 25 98	17 78 80 36 85	26 41 77 63 37	71 63 94 94 33
03 45 44 66 88	97 81 26 03 89	39 46 67 21 17	98 10 39 33 15	61 63 00 25 92
89 41 58 91 63	65 99 59 97 84	90 14 79 61 55	56 16 88 87 60	32 15 99 67 43
13 43 00 97 26	16 91 21 32 41	60 22 66 72 17	31 85 33 69 07	68 49 20 43 29
71 71 00 51 72	62 03 89 26 32	35 27 99 18 25	78 12 03 09 70	50 93 19 35 56
19 28 15 00 41	92 27 73 40 38	37 11 05 75 16	98 81 99 37 29	92 20 32 39 67
56 38 30 92 30	45 51 94 69 04	00 84 14 36 37	95 66 39 01 09	21 68 40 95 79
39 27 52 89 11	00 81 06 28 48	12 08 05 75 26	03 35 63 05 77	13 81 20 67 58
73 13 28 58 01	05 06 42 42 07	60 60 29 99 93	72 93 78 04 36	25 76 01 54 03
81 60 84 51 57	12 68 46 55 89	60 09 71 87 89	70 81 10 95 91	83 79 68 20 66
05 62 98 07 85	07 79 26 69 61	67 85 72 37 41	85 79 76 48 23	61 58 87 08 05
62 97 16 29 18	52 16 16 23 56	62 95 80 97 63	32 25 34 03 36	48 84 60 37 65
31 13 63 21 08	16 01 92 58 21	48 79 74 73 72	08 64 80 91 38	07 28 66 61 59
97 38 35 34 19	89 84 05 34 47	88 09 31 54 88	97 96 86 01 69	46 13 95 65 96
32 11 78 33 82	51 99 98 44 39	12 75 10 60 36	80 66 39 94 97	42 36 31 16 59
81 99 13 37 05	08 12 60 39 23	61 73 84 89 18	26 02 04 37 95	96 18 69 06 30
45 74 00 03 05	69 99 47 26 52	48 06 30 00 18	03 30 28 55 59	66 10 71 44 05
11 84 13 69 01	88 91 28 79 50	71 42 14 96 55	98 59 96 01 36	88 77 90 45 59
14 66 12 87 22	59 45 27 08 51	85 64 23 85 41	64 72 08 59 44	67 98 36 65 56
40 25 67 87 82	84 27 17 30 37	48 69 49 02 58	98 02 50 58 11	95 39 06 35 63
44 48 97 49 43	65 45 53 41 07	14 83 46 74 11	76 66 63 60 08	90 54 33 65 84
41 94 54 06 57	48 28 01 83 84	09 11 21 91 73	97 28 44 74 06	22 30 95 69 72
07 12 15 58 84	93 18 31 83 45	54 52 62 29 91	53 58 54 66 05	47 19 63 92 75
64 27 90 43 52	18 26 32 96 83	50 58 45 27 57	14 96 39 64 85	73 87 96 76 23
80 71 86 41 03	45 62 63 40 88	35 69 34 10 94	32 22 52 04 74	69 63 21 83 41
27 06 08 09 92	26 22 59 28 27	38 58 22 14 79	24 32 12 38 42	33 56 90 92 57
54 68 97 20 54	33 26 74 03 30	74 22 19 13 48	30 28 01 92 49	58 61 52 27 03
02 92 65 68 99	05 53 15 26 70	04 69 22 64 07	04 73 25 74 82	78 35 22 21 88
83 52 57 78 62	98 61 70 48 22	68 50 64 55 75	42 70 32 09 60	58 70 61 43 97
82 82 76 31 33	85 13 41 38 10	16 47 61 43 77	83 27 19 70 41	34 78 77 60 25
38 61 34 09 49	04 41 66 09 76	20 50 73 40 95	24 77 95 73 20	47 42 80 61 03
01 01 11 88 38	03 10 16 82 24	39 58 20 12 39	82 77 02 18 88	33 11 49 15 16
21 66 14 38 28	54 08 18 07 04	92 17 63 36 75	33 14 11 11 78	97 30 53 62 38
32 29 30 69 59	68 50 33 31 47	15 64 88 75 27	04 51 41 61 96	86 62 93 66 71
04 59 21 65 47	39 90 89 86 77	46 86 86 88 86	50 09 13 24 91	54 80 67 78 66
38 64 50 07 36	56 50 45 94 25	48 28 48 30 51	60 73 73 03 87	68 47 37 10 84
48 33 50 83 53	59 77 64 59 90	58 92 62 50 18	93 09 45 89 06	13 26 98 86 29

From Fisher and Yates; courtesy of Longman Group Ltd., London.

Random numbers (VI)

25 19 64 82 84	62 74 29 92 24	61 03 91 22 48	64 94 63 15 07	66 85 12 00 27
23 02 41 46 04	44 31 52 43 07	44 06 03 09 34	19 83 94 62 94	48 28 01 51 92
55 85 66 96 28	28 30 62 58 83	65 68 62 42 45	13 08 60 46 28	95 68 45 52 43
68 45 19 69 59	35 14 82 56 80	22 06 52 26 39	59 78 98 76 14	36 09 03 01 86
69 31 46 29 85	18 88 26 95 54	01 02 14 03 05	48 00 26 43 85	33 93 81 45 95
37 31 61 28 98	94 61 47 03 10	67 80 84 41 26	88 84 59 69 14	77 32 82 81 89
66 42 19 24 94	13 13 38 69 96	76 69 76 24 13	43 83 10 13 24	18 32 84 85 04
33 65 78 12 35	91 59 11 38 44	23 31 48 75 74	05 30 08 46 32	90 04 93 56 16
76 32 06 19 35	22 95 30 19 29	57 74 43 20 90	20 25 36 70 69	38 32 11 01 01
43 33 42 02 59	20 39 84 95 61	58 22 04 02 99	99 78 78 83 82	43 67 16 38 95
28 31 93 43 94	87 73 19 38 47	54 36 90 98 10	83 43 32 26 26	22 00 90 59 22
97 19 21 63 34	69 33 17 03 02	11 15 50 46 08	42 69 60 17 42	14 68 61 14 48
82 80 37 14 20	56 39 59 89 63	33 90 38 44 50	78 22 87 10 88	06 58 87 39 67
03 68 03 13 60	64 13 09 37 11	86 02 57 41 99	31 66 60 65 64	03 03 02 58 97
65 16 58 11 01	98 78 80 63 23	07 37 66 20 56	20 96 06 79 80	33 39 40 49 42
24 65 58 57 04	18 62 85 28 24	26 45 17 82 76	39 65 01 73 91	50 37 49 38 73
02 72 64 07 75	85 66 48 38 73	75 10 96 59 31	48 78 58 08 88	72 08 54 57 17
79 16 78 63 99	43 61 00 66 42	76 26 71 14 33	33 86 76 71 66	37 85 05 56 07
04 75 14 93 39	68 52 16 83 34	64 09 44 62 58	48 32 72 26 95	32 67 35 49 71
40 64 64 57 60	97 00 12 91 33	22 14 73 01 11	83 97 68 95 65	67 77 80 98 87
06 27 07 34 26	01 52 48 69 57	19 17 53 55 96	02 41 03 89 33	86 85 73 02 32
62 40 03 87 10	96 88 22 46 94	35 56 60 94 20	60 73 04 84 98	96 45 18 47 07
00 98 48 18 97	91 51 63 27 95	74 25 84 03 07	88 29 04 79 84	03 71 13 78 26
50 64 19 18 91	98 55 83 46 09	49 66 41 12 45	41 49 36 83 43	53 75 35 13 39
38 54 52 25 78	01 98 00 89 85	86 12 22 89 25	10 10 71 19 45	88 84 77 00 07
46 86 80 97 78	65 12 64 64 70	58 41 05 49 08	68 68 88 54 00	81 61 61 80 41
90 72 92 93 10	09 12 81 93 63	69 30 02 04 26	92 36 48 69 45	91 99 08 07 65
66 21 41 77 60	99 35 72 61 22	52 40 74 67 29	97 50 71 39 79	57 82 14 88 06
87 05 46 52 76	89 96 34 22 37	27 11 57 04 19	57 93 08 35 69	07 51 19 92 66
46 90 61 03 06	89 85 33 22 80	34 89 12 29 37	44 71 38 40 37	15 49 55 51 08
11 88 53 06 09	81 83 33 98 29	91 27 59 43 09	70 72 51 49 73	35 97 25 83 41
11 05 92 06 97	68 82 34 08 83	25 40 58 40 64	56 42 78 54 06	60 96 96 12 82
33 94 24 20 28	62 42 07 12 63	34 39 02 92 31	80 61 68 44 19	09 92 14 73 49
24 89 74 75 61	61 02 73 36 85	67 28 50 49 85	37 79 95 02 66	73 19 76 28 13
15 19 74 67 23	61 38 93 73 68	76 23 15 58 20	35 36 82 82 59	01 33 48 17 66
05 64 12 70 88	80 58 35 06 88	73 48 27 39 43	43 40 13 35 45	55 10 54 38 50
57 49 36 44 06	74 93 55 39 26	27 70 98 76 68	78 36 26 24 06	43 24 56 40 80
77 82 96 96 97	60 42 17 18 48	16 34 92 19 52	98 84 48 42 92	83 19 06 77 78
24 10 70 06 51	59 62 37 95 42	53 67 14 95 29	84 65 43 07 30	77 54 00 15 42
50 00 07 78 23	49 54 36 85 14	18 50 54 18 82	23 79 80 71 37	60 62 95 40 30
44 37 76 21 96	37 03 08 98 64	90 85 59 43 64	17 79 96 52 35	21 05 22 59 30
90 57 55 17 47	53 26 79 20 38	69 90 58 64 03	33 48 32 91 54	68 44 90 24 25
50 74 64 67 42	95 28 12 73 23	32 54 98 64 94	82 17 18 17 14	55 10 61 64 29
44 04 70 22 02	84 31 64 64 08	52 55 04 24 29	91 95 43 81 14	66 13 18 47 44
32 74 61 64 73	21 46 51 44 77	72 48 92 00 05	83 59 89 65 06	53 76 70 58 78
75 73 51 70 49	12 53 67 51 54	38 10 11 67 73	22 32 61 43 75	31 61 22 21 11
76 18 36 16 34	16 28 25 82 98	64 26 70 54 87	49 48 55 11 39	94 25 20 80 85
00 17 37 71 81	64 21 91 15 82	81 04 14 52 11	39 07 30 60 77	39 18 27 85 68
54 95 57 55 04	12 77 40 70 14	79 86 61 57 50	52 49 41 73 46	05 63 34 92 33
69 99 95 54 63	44 37 33 53 17	38 06 58 37 93	47 10 62 31 28	63 59 40 40 32

From Fisher and Yates; courtesy of Longman Group Ltd., London.

References

AGNEW, A. D. Q. (1961). The ecology of *Juncus effusus* L. in North Wales. *J. Ecol.*, **49**, 83–102.

AGNEW, A. D. Q. and HAINES, R. W. (1960). Studies on the plant ecology of the Jazira of Central Iraq. *I. Bull. Col. Sci. Baghdad*, **5**, 41–60.

ALVIN, K. L. (1960). Observations on the lichen ecology of South Haven peninsula, Studland Heath, Dorset. *J. Ecol.*, **48**, 331–9.

ANDERSON, D. J. (1961*a*). The structure of some upland plant communities in Caernarvonshire. II—The pattern shown by *Vaccinium myrtillus* and *Calluna vulgaris*. *J. Ecol.*, **49**, 731–8.

ANDERSON, D. J. (1961*b*). The structure of some upland plant communities in Caernarvonshire. I—The pattern shown by *Pteridium aquilinum*. *J. Ecol.*, **49**, 369–77.

ARCHIBALD, E. E. A. (1948). Plant populations—I. A new application of Neyman's contagious distribution. *Ann. Bot., Lond.*, N.S., **12**, 221–35.

ARCHIBALD, E. E. A. (1950). Plant populations—II. The estimation of the number of individuals per unit area of species in heterogenous plant populations. *Ann. Bot., Lond.*, N.S., **14**, 7–21.

ASHBY, E. (1935). The quantitative analysis of vegetation. *Ann. Bot., Lond.*, **49**, 779–802.

ASHBY, E. (1948). Statistical ecology, a re-assessment. *Bot. Rev.*, **14**, 222–34.

ASHBY, E. and PIDGEON, I. M. (1942). A new quantitative method of analysis of plant communities. *Aust. J. Sci.*, **5**, 19.

ASPINALL, D. and MILTHORPE, F. L. (1959). An analysis of competition between barley and white persicaria. I. The effects on growth. *Ann. appl. Biol.*, **47**, 156–72.

AUSTIN, M. P. and ORLOCI, L. (1966). Geometric models in ecology. II. An evaluation of some ordination techniques. *J. Ecol.*, **54**, 217–27.

AUSTIN, M. P. (1968). Pattern in a *Zerna erecta* dominated community. *J. Ecol.*, **56**, 197–218.

AUSTIN, M. P. and GREIG-SMITH, P. (1968). The application of quantitative methods to vegetation survey. II. Some methodological problems of data from rain forest. *J. Ecol.*, **56**, 827–44.

BARCLAY-ESTRUP, P. and GIMINGHAM, C. H. (1969). The description and interpretation of cyclical processes in a heath community. I. Vegetational change in relation to the *Calluna* cycle. *J. Ecol.*, **57**, 737–58.

BARCLAY-ESTRUP, P. (1971). The description and interpretation of cyclical processes in a heath community. III. Micro-climate in relation to the *Calluna* cycle. *J. Ecol.*, **59**, 143–66.

BARKMAN, J. J. (1949). Le Fabronietum pusillae et quelques autres associations épiphytiques du Tessin. (Suisse Meridionale.) *Vegetatio*, **2**, 309.

BARNES, H. and STANBURY, F. A. (1951). A statistical study of plant distribution during the colonization and early development of vegetation on china clay residues. *J. Ecol.*, **39**, 171–81.

BEARD, J. S. (1949). The natural vegetation of the Windward and Leeward Islands. *Oxf. For. Mem.*, No. 21.

BEDFORD, DUKE OF, and PICKERING, S. U. (1919). *Science and Fruit growing*. Macmillan, London.

BESCHEL, R. E. (1961). Dating rock surfaces by lichen growth and its application to glaciology and physiography (Lichenometry). *Geology of the Arctic*. 1044–62. Univ. Toronto Press.

BILLINGS, W. D. and MARK, A. F. (1961). Interactions between alpine tundra vegetation and patterned ground in the mountains of southern New Zealand. *Ecology*, **42**, 18–31.

BILLINGS, W. D. and MOONEY, H. A. (1959). An apparent frost hummock-sorted polygon cycle in the alpine tundra of Wyoming. *Ecology*, **40**, 16–20.

BJORKMAN, O. and HOLMGREN, P. (1963). Adaptability of the photosynthetic apparatus to light intensity in ecotypes from exposed and shaded habitats. *Physiol. Plant.*, **16**, 889–914.

BLACK, J. N. (1955). The influence of depth of sowing and temperature on pre-emergence weight changes in subterranean clover (*Trifolium subterraneum* L.) *Aust. J. agric. Res.*, **6**, 203.

BLACK, J. N. (1956). The influence of seed size and depth of sowing on pre-emergence and early vegetative growth of subterranean clover (*Trifolium subterraneum* L.) *Aust. J. agric. Res.*, **7**, 98–109.

BLACK, J. N. (1958). Competition between plants of different initial seed sizes in swards of subterranean clover (*Trifolium subterraneum* L.) with particular reference to leaf area and the light microclimate. *Aust. J. agric. Res.*, **9**, 299–318.

BLACK, J. N. (1960). The significance of petiole length, leaf area, and light interception in competition between strains of subterranean clover (*Trifolium subterraneum* L.) grown in swards. *Aust. J. agric. Res.*, **11**, 277–91.

BLACKMAN, G. E. (1935). A study by statistical methods of the distribution of species in grassland associations. *Ann. Bot., Lond.*, **49**, 749–78.

BLACKMAN, G. E. (1942). Statistical and ecological studies in the distribution of species in plant communities. I. Dispersion as a factor in the study of changes in plant populations. *Ann. Bot., Lond.*, N.S. **6**, 351–70.

BONNER, J. (1950). The role of toxic substances in the interactions of higher plants. *Bot. Rev.*, **16**, 51–65.

BRAUN-BLANQUET, J. (1927). *Pflanzensoziologie*. Springer, Wien.

BRAUN-BLANQUET, J. (1932). *Plant Sociology*.

BRAUN-BLANQUET, J. (1951). *Pflanzensoziologie*, 2nd edn. Springer, Wien.

BRAY, R. J. and CURTIS, J. T. (1957). An ordination of the upland forest communities of southern Wisconsin. *Ecol. Monogr.*, **27**, 325–49.

BRERETON, A. J. (1971). The structure of the species populations in the initial stages of salt-marsh succession. *J. Ecol.*, **59**, 321–38.

BROWN, R. T. and CURTIS, T. T. (1952). The upland conifer-hardwood forests of northern Wisconsin. *Ecol. Monogr.*, **22**, 217–34.

CAIN, S. A. (1938). The species-area curve. *Amer. Midl. Nat.*, **19**, 573.

CAIN, S. A. (1952). Concerning certain phytosociological concepts. *Ecol. Monogr.*, **2**, 475.

CAIN, S. A. (1954). Studies on virgin hardwood forest. II.—A comparison of quadrat sizes in a quantitative phytosociological study of Nash's Woods, Posey County, Indiana. *Amer. Midl. Nat.*, **15**, 529.

CALDWELL, P. A. (1957). The spatial development of *Spartina* colonies growing without competition. *Ann. Bot., Lond.*, **21**, 203–16.

CAVERS, P. B. and HARPER, J. L. (1966). Germination polymorphism in *Rumex crispus* and *Rumex obtusifolius*. *J. Ecol.*, **54**, 367–82.

CHADWICK, M. J. (1960). *Nardus stricta*, a weed of hill grazings. *The Biology of Weeds. Blackwell Sci. Pub.*, Oxford.

CHAPAS, L. C. (1959). Computer languages for model-building. In, '*The use of models in Agricultural and Biological research.*' Grassland. Res. Inst. Hurley. pp. 72–86.

CHRISTIAN, C. S. and PERRY, R. A. (1953). The systematic description of plant communities by the use of symbols. *J. Ecol.*, **41**, 100–5.

CLAPHAM, A. R. (1932). The form of the observational unit in quantitative ecology. *J. Ecol.*, **20**, 192–7.

CLAPHAM, A. R. (1936). Overdispersion in grassland communities and the use of statistical methods in plant ecology. *J. Ecol.*, **24**, 232–51.

CLEMENTS, F. E. (1904). Development and structure of vegetation. *Rep. Bot. Surv. Nebra.*, **7**.

CLEMENTS, F. E. (1916). Plant succession. An analysis of the development of vegetation. *Carnegie Inst. Washington.* No. 242.

COOK, M. T. (1921). Wilting caused by walnut trees. *Phytopathology*, **11**, 346.

COOPER, C. F. (1960). Changes in vegetation. Structure and growth of south western pine forests since white settlement. *Ecol. Monogr.*, **30**, 129–64.

COOPER, C. F. (1961). Pattern in ponderosa pine forests. *Ecology*, **42**, 493–9.

COOPER, W. S. (1923). The recent ecological history of Glacier Bay, Alaska. *Ecology*, **6**, 197.

COOPER, W. S. (1926). The fundamentals of vegetational change. *Ecology*, **7**, 391–413.

COOPER, W. S. (1931). A third expedition to Glacier Bay, Alaska. *Ecology*, **12**, 61–95.

COOPER, W. S. (1937). The problem of Glacier Bay, Alaska. A study of glacier variations. *Geogr. Rev.*, **27**, 37–62.

COOPER, W. S. (1939). A fourth expedition to Glacier Bay, Alaska. *Ecology*, **20**, 130–59.

COTTAM, G. (1949). The phytosociology of an oak wood in south-western Wisconsin. *Ecology*, **30**, 271–87.

COWLES, H. C. (1899). The ecological relations of the vegetation on the sand dunes of Lake Michigan. *Bot. Gaz.*, **27**, 95–117, 167–202, 281–308, 361–91.

COWLES, H. C. (1901). The physiographic ecology of Chicago and vicinity. A study of the origin, development and classification of plant societies. *Bot. Gaz.*, **31**, 73.

CROCKER, R. L. and MAJOR, J. (1955). Soil development in relation to vegetation and surface age at Glacier Bay, Alaska. *J. Ecol.*, **43**, 427–48.

CURTIS, J. T. (1947). The paloverde forest type near Gonivaves, Haiti, and its relation to the surrounding vegetation. *Caribb. Forester*, **8**, 1–26.

CURTIS, J. T. and MCINTOSH, R. P. (1950). The inter-relations of certain analytic and synthetic phytosociological characters. *Ecology*, **31**, 434–55.

CURTIS, J. T. and MCINTOSH, R. P. (1951). An upland forest continuum in the prairies-forest border region of Wisconsin. *Ecology*, **32**, 476–96.

DAGNELIE, P. (1960). Contribution à l'étude des communautés végétales par l'analyse factorielle. *Bull. Serv. Carte. Phytogeogr. Sér. B.*, **5**, 7–71, 93–195.

DAHL, E. and HADAČ, E. (1941). Strandgesellschaften der Insel Ostroy im Oslofjord. *Nyt Mag. Naturvid.*, **82**, 251.

DAHL, E. and HADAČ, E. (1949). Homogeneity of plant communities. *Studia bot. Čechosl.*, **10**, 159–76.

DALE, M. B. (1970). Systems analysis and ecology. *Ecology*, **51**, 2–16.

DAVID, F. N. and MOORE, P. G. (1954). Notes on contagious distributions in plant populations. *Ann. Bot., Lond.*, N.S. **18**, 47–53.

DAVIS, E. F. (1928). The toxic principle of *Juglans nigra* as identified with synthetic juglone, and its toxic effects on tomato and alfalfa plants. *Amer. J. Bot.*, **15**, 620.

DAVIS, T. A. W. and RICHARDS, P. W. (1933–4). The vegetation of Moraballi Creek, British Guiana, an ecological study of a limited area of Tropical Rain Forest. Parts I and II. *J. Ecol.*, **21**, 350–84; **22**, 106–55.

DECANDOLLE, A. P. (1832). *Physiologie végétale*, T. **3**, 1474–5. Paris.

DE VRIES, D. M. (1953). Objective combinations of species *Acta bot. neerl.*, **1**, 497–9.

DICE, L. R. (1952). Measure of the spacing between individuals within a population. *Contr. Lab. vertebr. Biol. Univ. Mich.*, **55**, 1–23.

DONALD, C. M. (1951). *Aust. J. agric. Res.*, **2**, 355.

DONALD, C. M. (1961). *Symp. Soc. exp. Biol.*, **51**, 282–313.

DUNCAN, W. G., LOOMIS, R. S., WILLIAMS, W. A. and HANAU, R. (1967). A model for simulating photosynthesis in plant communities. *Hilgardia*, **38**, 181–205.

DU RIETZ, G. E. (1921). Zur methodlogischen Grundlage der Modernen Pflanzensoziologie. *Akad. Abh. Uppsala.*

EMMETT, H. E. G. and ASHBY, E. (1934). Some observations between H-ion concentration of the soil and plant succession. *Ann. Bot., Lond.*, **68**, 869.

EVANS, F. C. (1952). The influence of size of quadrat on the distributional patterns of plant populations. *Contr. Lab. vertebr. Biol. Univ. Mich.*, **54**, 1–15.

FIELD, W. O. (1947). Glacier recession in Muir Inlet, Glacier Bay, Alaska. *Geogr. Rev.*, **37**, 369–99.

FISHER, R. A. and YATES, F. (1948). *Statistical tables for biological, agricultural and medical research.* Oliver and Boyd, London.

FLENLEY, J. R. (1969). The vegetation of the Wabag region, New Guinea highlands: A numerical study. *J. Ecol.*, **57**, 465–90.

GIMINGHAM, C. H., PRITCHARD, N. M. and CORMACK, R. M. (1966). Interpretation of a vegetational mosaic on limestone in the island of Gotland. *J. Ecol.*, **54**, 481–502.

GITTINS, R. (1965). Multivariate approaches to a limestone grassland community. III. A comparative study of ordination and association—analysis. *J. Ecol.*, **53**, 411–25.

GLEASON, H. A. (1917). The structure and development of the plant association. *Bull. Torrey bot. Club*, **43**, 463–81.

GLEASON, H. A. (1920). Some applications of the quadrat method. *Bull. Torrey bot. Club*, **47**, 21–33.

GLEASON, H. A. (1922). The vegetational history of the Middle West. *Ann. Ass. Amer. Geogr.*, **12**, 39–85.

GLEASON, H. A. (1926). The individualistic concept of the plant association. *Bull. Torrey bot. Club*, **53**, 7–26.

GODWIN, H. and CONWAY, V. M. (1939). The ecology of a raised bog near Tregaron, Cardiganshire. *J. Ecol.*, **27**, 313–59.

GOLDEN, J. T. (1965). *Fortran IV Programming and Computing*. Prentice-Hall, Inc., Englewood Cliffs, N. J.

GOODALL, D. W. (1952*a*). Some considerations in the use of point quadrats for the analysis of vegetation. *Aust. J. sci. Res.*, **5**, 1–41.

GOODALL, D. W. (1952*b*). Quantitative aspects of plant distribution. *Biol. Rev.*, **27**, 194–245.

GOODALL, D. W. (1953). Objective methods for the classification of vegetation. I.—The use of positive interspecific correlation. *Aust. J. Bot.*, **1**, 39–63.

GOODALL, D. W. (1954*a*). Vegetational classification and vegetational continua. *Angew. Pflanzensoziologie*, **1**, 168–82.

GOODALL, D. W. (1945*b*). Objective methods for the classification of vegetation. 3—An essay in the use of factor analysis. *Aust. J. Bot.*, **2**, 304–24.

GOODWIN, R. H. and TAVES, C. (1950). The effect of coumarin derivations on the growth of *Avena* roots. *Amer. J. Bot.*, **37**, 224–31.

GORHAM, E. (1953). Chemical studies on the soils and vegetation of water-logged habitats in the English Lake District. *J. Ecol.*, **41**, 345–60.

GRAY, R. and BONNER, J. (1948). Structure, determination and synthesis of a plant growth inhibitor, 3-acetyl-6-methoxybenzaldehyde found in the leaves of *Encelia farinosa*. *J. Amer. chem. Soc.*, **70**, 1249–53.

GRAY R. and BONNER J. (1948). An inhibitor of plant growth from the leaves of *Encelia farinosa*.

GREIG-SMITH, P. (1952*a*). The use of random and contiguous quadrats in the study of the structure of plant communities. *Ann. Bot., Lond.*, N.S. **16**, 293–316.

GREIG-SMITH, P. (1952*b*). Ecological observations on degraded and secondary forest in Trinidad, British West Indies. *J. Ecol.*, **40**, 316–30.

GREIG-SMITH, P. (1957). *Quantitative Plant Ecology*. Butterworths, London.

GREIG-SMITH, P. (1961*a*). Data on pattern within plant communities. I.—The analysis of pattern. *J. Ecol.*, **49**, 695–702.

GREIG-SMITH, P. (1961*b*). Data on pattern within plant communities. II. *Ammophila arenaria* (L.) Link. *J. Ecol.*, **49**, 703–8.

GREIG-SMITH, P., AUSTIN, M. P. and WHITMORE, T. C. (1967). The application of quantitative methods to vegetation survey. I. Association—analysis and principal component ordination of rain-forest. *J. Ecol.*, **55**, 483–503.

HALL, J. B. (1967). Some aspects of the ecology of *Brachypodium pinnatum*. Ph.D. thesis, Imperial College, London.

HALL, J. B. (1971). Pattern in a chalk grassland community. *J. Ecol.*, **59**, 749–762.

HARDWICK, R. C. (1969). The pea crop. In *The use of models in Agricultural and Biological research. Grassland Res. Inst. Hurley.* pp. 24–34.

HARPER, J. L. (1957). The ecological significance of dormancy and its importance in weed control. *Proc. 4th int. Congr. Pl. Prot.*, pp. 415–20.

HARPER, J. L. (1960). Factors controlling plant numbers. *The Biology of Weeds.* (Br. Ecol. Soc. Symp.) Blackwell Sci. Pub., Oxford. pp. 119–32.

HARPER, J. L. (1961). Approaches to the study of plant competition. *Symp. Soc. exp. Biol.*, **15**, 1–39.

HARPER, J. L. and CHANCELLOR, A. P. (1959). The comparative biology of closely related species living in the same area. IV. *Rumex:* Interference between individuals in populations of one and two species. *J. Ecol.*, **47**, 679–95.

HARPER, J. L. and CLATWORTHY, J. N. (1963). The comparative biology of closely related species. VI. Analysis of the growth of *Trifolium repens* and *T. fragiferum* in pure and mixed populations. *J. exp. Bot.*, **14**, 172–90.

HARPER, J. L., LANDRAGIN, P. A. and LUDWIG, J. W. (1955). The influence of environment on seed and seedling mortality. I. The influence of time of planting on the germination of maize. *New Phytol.*, **54**, 107–18.

HARPER, J. L. and MCNAUGHTON, J. H. (1962). The comparative biology of closely related species living in the same area. VII. Interference between individuals in pure and mixed populations of *Papaver* species. *New Phytol.*, **61**, 175–88.

HARPER, J. L. and SAGAR, G. R. (1953). Some aspects of the Ecology of Buttercups in permanent grassland. *Proc. Br. Weed Control Conf. 1953*, pp. 256–65.

HARPER, J. L., WILLIAMS, J. T. and SAGER, G. R. (1965). The behaviour of seeds in soil. I. The heterogeneity of soil surfaces and its role in determining the establishment of plants from seed. *J. Ecol.*, **53**, 273–86.

HARRIS, G. P. (1972). The ecology of corticolous lichens. III. A simulation model of productivity as a function of light intensity and water availability. *J. Ecol.*, **60**, 19–40.

HARRIS, G. P. (1974). The diel and annual cycles of net plankton photosynthesis in Lake Ontario. *J. Fish. Res. Bd. Can.* (In press.)

HARRISON, C. M. (1970). The phytosociology of certain English heathland communities. *J. Ecol.*, **58**, 573–89.

HESLOP-HARRISON, J. (1964). Forty years of Genecology. In *Adv. Ecol. Res.*, **2**. Academic Press, London.

HOLLING, C. S. (1966). The strategy of building models of complex ecological systems. In: *Systems analysis in ecology*. K. E. F. WATT, ed. Academic Press N.Y. pp. 195–214.

HOPE-SIMPSON, J. F. (1940). On the errors in the ordinary use of subjective frequency estimations in grassland. *J. Ecol.*, **28**, 193–209.

HOPKINS, B. (1954). A new method for determining the type of distribution of plant individuals. *Ann. Bot., Lond.*, N.S. **18**, 213–27.

HOPKINS, B. (1955). The species-area relations of plant communities. *J. Ecol.*, **43**, 409–26.

HOPKINS, B. (1957). The concept of the minimal area. *J. Ecol.*, **45**, 441–9.

HOPKINS, B. (1968). Vegetation of the Olokemeji forest reserve, Nigeria. V. The vegetation on the savanna site with special reference to its seasonal changes. *J. Ecol.*, **56**, 97–115.

HUBBELL, S. P. (1971). Of sowbugs and systems: The ecological bioenergetics of a terrestrial isopod. In: *Systems analysis and Simulation Ecology*. Academic Press N.Y.

IVIMEY-COOK, R. B., PROCTOR, M. C. F. and WIGSTON, D. L. (1969). On the problem of the 'R/Q' terminology in multivariate analyses of biological data. *J. Ecol.*, **57**, 673–5.

JONES, E. W. (1955–6). Ecological studies on the rain forest of southern Nigeria. IV.—The plateau forest of the Okomu Forest Reserve. *J. Ecol.*, **43**, 564–94; **44**, 83–117.

JONES, L. R. and MORSE, W. J. (1903). The shrubby cinquefoil as a weed. *Vt. Agr. Emp. Sta. Ann. Rep.*, **61**, 173–90.

KENDALL, M. G. (1957). *A course in multivariate analysis*. London.

KERSHAW K. A. (1957). The use of cover and frequency in the detection of pattern in plant communities. *Ecology* **38**, 291–9.

KERSHAW, K. A. (1958, 1959). An investigation of the structure of a grassland community. I. The pattern of *Agrostis tenuis*. *J. Ecol.*, **46**, 571–92; II. The pattern of *Dactylis glomerata*, *Lolium perenne* and *Trifolium repens*; III. Discussion and conclusions. *J. Ecol.*, **47**, 31–53.

KERSHAW, K. A. (1960a). The detection of pattern and association. *J. Ecol.*, **48**, 233–42.

KERSHAW, K. A. (1960b). Cyclic and pattern phenomena as exhibited by *Alchemilla alpina*. *J. Ecol.*, **48**, 443–53.

KERSHAW, K. A. (1961). Association and co-variance analysis of plant communities. *J. Ecol.*, **49**, 643–54.

KERSHAW, K. A. (1962a). Quantitative ecological studies from Landmannahellir, Iceland. I. *Eriophorum angustifolium*. *J. Ecol.*, **50**, 171–9.

KERSHAW, K. A. (1962b). Quantitative ecological studies from Landmannahellir, Iceland. II. The rhizome behaviour of *Carex bigelowii* and *Calamagrostis neglecta*. *J. Ecol.*, **50**, 171–9.

KERSHAW, K. A. (1962c). Quantitative ecological studies from Landmannahellir, Iceland. III. The variation of performance of *Carex bigelowii*. *J. Ecol.*, **50**, 393–9.

KERSHAW, K. A. (1963). Pattern in vegetation and its causality. *Ecology*, **44**, 377–88.

KERSHAW, K. A. (1968). Classification and ordination of Nigerian savanna vegetation. *J. Ecol.*, **56**, 467–82.

KERSHAW, K. A. and HARRIS, G. P. (1969). Simulation studies and ecology: A simple defined system model. *Statistical Ecology*, **3**, 1–21. Penn. State Univ. Press.

KERSHAW, K. A. and HARRIS, G. P. (1969). Simulation studies and ecology: use of the model. *Statistical Ecology*, **3**, 23–42. Penn. State Univ. Press.

KERSHAW, K. A. and ROUSE, W. R. (1971a). Studies on lichen-dominated systems. I. The water relations of *Cladonia alpestris* in spruce-lichen woodland in northern Ontario. *Can. J. Bot.*, **49**, 1389–99.

KERSHAW, K. A. and ROUSE, W. R. (1971b). Studies on lichen-dominated systems. II. The growth pattern of *Cladonia alpestris* and *Cladonia rangiferina*. *Can. J. Bot.*, **49**, 1401–10.

KERSHAW, K. A. and SHEPHERD, R. (1972). Computer display graphics for principal component analysis and vegetation ordination studies. *Can. J. Bot.*, **50**, 2239–50.

KERSHAW, K. A. and TALLIS, J. H. (1958). Pattern in the high level *Juncus squarrosus* community. *J. Ecol.*, **46**, 739–48.

LAWRENCE, D. B. (1953). Development of vegetation and soil in south-eastern Alaska, with special reference to the accumulation of nitrogen. Final Report ONR, Project NR, pp. 160–83.

LAWRENCE, D. B. (1958). Glaciers and vegetation in south-eastern Alaska. *Amer. Scient.*, **46**, 89–122.

LAZENBY, A. (1955). Germination and establishment of *Juncus effusus* L. The effect of different companion species and of variation in soil and fertility conditions. *J. Ecol.*, **43**, 103–19.

LIBBERT, E. and LÜBKE, H. (1957). Physiologische Wirkungen des Scopoletins, I. Der Einfluss des Scopoletins auf die Samenkeimung. *Flora, Jena*, **145**, 256–63; (1958) II. Der Einfluss des Scopoletins auf des Wurzelwachstum. *Flora, Jena*, **146**, 228–39.

LOOMIS, R. S., WILLIAMS, W. A. and DUNCAN, W. G. (1967). Community Architecture and the productivity of terrestrial plant communities. In *Harvesting the Sun Photosynthesis in Plant Life*. Academic Press, N.Y. pp. 291–308.

MARTIN, P. (1956). Qualitative und quantitative Untersuchungen über die Ausscheidung organischer Verbindungen aus den Keimwurzeln des Hafers. (*Avena sativa* L.) *Naturwissenschaften*, **43**, 227–8.

MARTIN, P. (1957). Die Abgabe von organischen Verbindungen, insbesondere von Scopoletin, aus den Keinwurzeln des Hafers. *Z. Bot.*, **45**, 475–506.

MARTIN, P. (1958). Einfluss der Kulturfiltrate von Mikroorganismen auf die Abgabe von Scopoletin aus den Keinwurseln des Hafers. (*Avena sativa* L.) *Arch. Mikrobiol.*, **29**, 154–68.

MASSEY, A. B. (1925). Antagonism of the walnuts (*Juglans nigra* L. and *J. cinerea* L.) in certain plant associations. *Phytopathology*, **15**, 773–84.

MAURER, W. D. (1968). *Programming: an introduction to computer languages and techniques*. Holden-Day Inc. San Francisco.

MCCRACKEN, D. (1965). *A guide to Fortran IV programming*. John Wiley and Sons. N.Y.

MCKELL, C. M., PERRIER, E. R. and STEBBINS, G. L. (1960). Responses of two subspecies of orchard grass (Dactylis glomerata subspecies *lusitanica* and *judaica*) to increasing soil moisture stress. *Ecology*, **41**, 772–8.

MCQUEEN, D. J. (1971). A components study of competition in two cellular slime mold species: *Dictyostelium discoideum* and *Polysphondylium pallidum*. *Can. J. Zool.*, **49**, 1163–77.

MCVEAN, D. N. and RATCLIFFE, D. A. (1962). *Plant communities of the Scottish Highlands*. H.M.S.O., London.

MICHELMORE, A. P. G. (1934). Vegetational succession and regional surveys with special reference to tropical Africa. *J. Ecol.*, **22**, 313–17.

MOONEY, H. A. and BILLINGS, W. D. (1961). Comparative physiological ecology of arctic and alpine populations of *Oxyria digyna*. *Ecol. Monogr.*, **31**, 1–29.

MOORE, J. J. (1962). The Braun-Blanquet system. A reassessment. *J. Ecol.*, **50**, 761–9.

MOORE, P. G. (1953). A test for non-randomness in plant populations. *Ann. Bot., Lond.*, N.S. **17**, 57–62.

MOORE, P. G. (1954). Spacing in plant populations. *Ecology*, **35**, 222–7.

NEEL, R. B. and OLSON, J. S. (1962). Use of analog computers for simulating the movement of isotopes in ecological systems. Oak Ridge Nat. Lab. Report ORNL—3172.

NEWNHAM, R. M. (1968). A classification of climate by principal component analysis and its relationship to tree species distribution. *Forest Sci.* **14**, 254–64.

NICHOLSON, L. A. and ROBERTSON, R. A. (1958). Some observations on the ecology on an upland grazing in north-east Scotland with particular reference to Callunetum. *J. Ecol.*, **46**, 239–70.

ODUM, E. P. (1953). *Fundamentals of ecology*. W. B. Saunders, Philadelphia.

ODUM, E. P. (1971). *Fundamentals of ecology*. W. B. Saunders, Philadelphia.

ODUM, H. T. (1956). Primary production in flowing waters. *Limnol. Oceanogr.*, **1**, 102–17.

ODUM, H. T. (1957). Tropic structure and productivity of Silver Springs, Florida. *Ecol. Monogr.*, **27**, 55–112.

ODUM, H. T. (1960). Ecological potential and analogue circuits for the ecosystem. *Amer. Sci.*, **48**, 1–8.

OLSON, J. S. (1958). Rates of succession and soil changes on Southern Lake, Michigan, sand dunes. *Bot. Gaz.*, **119**, 125–70.

OLSON, J. S. (1964). Gross and net production of terrestrial vegetation. *J. Ecol. Jub. Symp. Sup.*, pp. 99–118.

ORLOCI, L. (1966). Geometric models in ecology. I. The theory and application of some ordination methods. *J. Ecol.*, **54**, 193–215.

OSWALD, H. (1923). *Die vegetation des Hochmoores Komosse.* Akad. Abh. Uppsala.

OVINGTON, J. D. (1953). A study of invasion by *Holcus mollis* L. *J. Ecol.*, **41**, 35–52.

PATTEN, B. C. (1971). *Systems Analysis and Simulation in Ecology.* Academic Press. N.Y.

PEARSON, G. A. (1942). Herbaceous vegetation a factor in natural regeneration of Ponderosa Pine in the south-west. *Ecol. Monogr.*, **12**, 315–38.

PÉNZES, A. (1958). A survival of stoloniferous plant colonies (polycormons) of a relict character. *Biologia, Sav.*, **13**. 253–64.

PÉNZES, A. (1960). Über die Morphologie Dynamik und Zönologisch Volle der Sprosskolonien—bildenden Pflanzen (Polycormone). *Geobotanika*, **6**, 501–515.

PERRY, G. S. (1932). Some tree antagonisms. *Penn. Acad. Sci. Proc.*, **6** 136–41.

PHILLIPS J. (1934–5). Succession, development, the climax and the complex organism. An analysis of concepts. Parts I and II. *J. Ecol.*, **22**, 559–71; **23**, 210–46; **23**, 488–508.

PHILLIPS, M. E. (1954a). Studies in the quantitative morphology and ecology of *Eriophorum angustifolium* Roth. II. Competition and dispersion. *J. Ecol.*, **42**, 187–210.

PHILLIPS, M. E. (1954b). Studies in the quantitative morphology and ecology of *Eriophorum angustifolium* Roth. III. The leafy shoot. *New Phytol.*, **53**, 312–43.

PIGOTT, C. D. (1956). The vegetation of upper Teesdale in the north Pennines. *J. Ecol.*, **44**, 545–86.

POORE, M. E. D. (1955–6). The use of phytosociological methods in ecological investigations. I. The Braun-Blanquet System. *J. Ecol.*, **43**, 226–44; II. Practical issues involved in an attempt to apply the Braun-Blanquet system. *J. Ecol.*, **43**, 245–69; III. Practical applications. *J. Ecol.*, **43**, 606–51; IV. General discussion of phytosociological problems. *J. Ecol.*, **44**, 28–50.

POORE, M. E. D. (1968). Studies in Malaysian rain forest. I. The forests on triassic sediments in Jengka forest reserve. *J. Ecol.*, **56**, 143–96.

POPAY, A. I. and ROBERTS, E. H. (1970a). Factors involved in the dormancy and germination of *Capsella bursa-pastoris* (L.) Medik. and *Senecio vulgaris* L. *J. Ecol.*, **58**, 103–22.

POPAY, A. I. and ROBERTS, E. H. (1970b). Ecology of *Capsella bursa-pastoris* (L.) Medik. and *Senecio vulgaris* L. in relation to germination behaviour. *J. Ecol.*, **58**, 123–39.

RADFORD, P. J. (1969). Some considerations governing the choice of a suitable simulation language. In, '*The use of models in Agricultural and Biological research*'. pp. 87–106. Grassland Res. Inst. Hurley.

RAUNKIAER, C. (1909). Formationsundersogelse og Formationsstatisk. *Bot. Tidsskr.*, **30**, 20–132.

RAUNKIAER, C. (1928). Dominansareal, Artstaethed og Formationsdominanter. Kgl. Danske Videnskabernes Selskab. *Biol. Medd. Kbh.*, **7**, 1.

RAUNKIAER, C. (1934). *The Life Forms of Plants and Statistical Plant Geography.* Translated by Carter, Fausboll and Tansley; Oxford Univ. Press.

RICHARDS, P. W. (1940). The recording of structure, life forms and flora of Tropical forest communities as a basis for their classification. *J. Ecol.*, **28**, 224–39.

RICHARDS, P. W. (1952). *The Tropical Rain Forest*. Univ. Press. Cambridge.

RICHARDS, P. W. (1956). Ecological observations on the rain forest of Mount Dulit, Sarawak. Parts I and II. *J. Ecol.*, **24**, 1–37; **24**, 340–60.

RICHARDS, P. W., TANSLEY, A. G. and WATT, A. S. (1940). The recording of structure, life form and flora of tropical forest communities as a basis for their classification. *J. Ecol.*, **28**, 224–39.

RUSSEL, E. W. (1961). *Soil Conditions and Plant Growth*. 9th edn., London.

SELLECK, G. W. (1960). The climax concept. *Bot. Rev.*, **26**, 534–45.

SKELLAM, J. G. (1952). Studies in statistical ecology. I. Spatial pattern. *Biometrica*, **39**, 346–62.

SNAYDON, R. W. (1962). Micro-distribution of *Trifolium repens* and its relation to soil factors. *J. Ecol.*, **50**, 133–43.

SNEDECOR, G. W. (1946). *Statistical Methods*. Iowa State College Press.

SOKAL, R. R. and SNEATH, H. A. (1963). *Principles of numerical taxonomy*. W. H. Freeman. San Francisco.

STOKES, P. (1965). Temperature and seed dormancy. *Handb. Pfl. Physiol.*, **15(2)**, 746–803.

SVEDBURG, T. (1922). Ettbidrag till de statiska metodernas användning inom vöxtbiologien. *Svensk bot. Tidskr.*, **16**, 1–8.

TANSLEY, A. G. (1916). The development of vegetation. *J. Ecol.*, **4**, 198–204.

TANSLEY, A. G. (1920). The classification of vegetation and the concept of development. *J. Ecol.*, **8**, 118–49.

THOMPSON, H. R. (1956). Distribution of distance to nth neighbour in a population of randomly distributed individuals. *Ecology*, **37**, 391–4.

THOMPSON, H. R. (1958). The statistical study of plant distribution patterns using a grid of quadrats. *Aust. J. Bot.*, **6**, 322–42.

THOMSON, G. W. (1952). Measures of plant aggregation based on contagious distributions. *Contr. Lab. Vertebr. Biol. Univ. Mich.*, **53**, 1–16.

THURSTON, J. M. (1959). *The Biology of Weeds*. Blackwell. Oxford. pp. 69–82.

TOOLE, E. H., HENDRICKS, S. B. and BORTHWICK, H. A. and TOOLE, V. K. (1956). Physiology of seed germination. *A. Rev. Physiol.*, **7**, 299–324.

VASILEVICH, V. I. (1961). Association between species and the structures of a phytocoenosis. *Dokl. obsch. Sobr. Ak. Nauk, S.S.S.R.*, **189**, 1001–4.

VEGIS, A. (1963). Climatic control of germination, bud break, and dormancy. *Environmental Control of Plant Growth*, pp. 265–87. N.Y.

VOLLENWEIDER, R. A. (1969). Calculation models of photosynthesis—depth curves and some implications regarding day rate estimates in primary production measurements. In: *Primary productivity in Aquatic Environments*. Univ. California Press. Berkeley and Los Angeles.

WAGGONER, P. E. and REIFSNYDER, W. E. (1968). Simulation of the temperature, humidity and evaporation profiles in a leaf canopy. *J. Appl. Met.*, **7**, 400–9.

WARD, S. D. (1970). The phytosociology of *Calluna-Arctostaphylos* heaths in Scotland and Scandinavia. *J. Ecol.*, **58**, 847–63.

WARMING, E. (1896). Lehrbuch der Oekologischen. *Pflanzengeographie*.

WARREN WILSON, J. (1952). Vegetational patterns associated with soil movement on Jan Mayen Island. *J. Ecol.*, **40**, 249–64.

WATT, A. S. (1919). On the cause of failure of natural regeneration in British oak woods. *J. Ecol.*, **17**, 173–203.

WATT, A. S. (1925). On the ecology of British beechwoods with special reference to their regeneration. II. The development and structure of beech communities on the South Downs. *J. Ecol.*, **13**, 27–73.

WATT, A. S. (1940). Studies in the Ecology of Breckland. IV. The Grass Heath. *J. Ecol.*, **28**, 42–70.

WATT, A. S. (1947*a*). Pattern and process in the plant community. *J. Ecol.*, **35**, 1–22.

WATT, A. S. (1947*b*). Contributions to the ecology of bracken. IV. The structure of the community. *New Phytol.*, **46**, 97–121.

WATT, A. S. (1955). Bracken versus heather, a study in plant sociology. *J. Ecol.*, **43**, 490–506.

WATT, A. S. (1962). The effect of excluding rabbits from grassland. A (Xerobrometum) in Breckland. 1936–60. *J. Ecol.*, **50**, 181–98.

WATT, A. S. and FRASER, G. K. (1933). Tree roots in the field layer. *J. Ecol.*, **21**, 404–14.

WEATHERELL, J. (1957). The use of nurse species in the afforestation of upland heaths. *Quat. J. For.*, LI, 1–7.

WEBB, D. A. (1954). Is the classification of plant communities either possible or desirable? *Saer. bot. Tidssk.*, **51**, 362–70.

WELCH, J. R. (1960). Observations on deciduous woodland in the eastern province of Tanganyika. *J. Ecol.*, **48**, 557–73.

WENT, F. W. (1942). The dependence of certain annual plants on shrubs in South California deserts. *Bull. Torrey bot. Cl.*, **69**, 100–14.

WENT, F. W. (1957). *The Experimental Control of Plant Growth.* Chronica Botanica Co., Waltham.

WERNHAM, C. C. (1951). Cold testing of corn. A chronological and critical review. *Pa. St. Coll. Agric. Exp. Sta. Progr. Report No. 47.*

WHITAKER, R. A. (1953). A consideration of climax theory: The climax as a population and pattern. *Ecol. Monogr.*, **23**, 41–78.

WHITE, J. and HARPER, J. L. (1970). Correlated changes in plant size and number in plant populations. *J. Ecol.*, **58**, 467–85.

WHITFORD, P. B. (1949). Distribution of woodland plants in relation to succession and clonal growth. *Ecology*, **30**, 199–208.

WILLIAMS, W. T. and LAMBERT, J. M. (1959). Multivariate methods in plant ecology. I. Association analysis in plant communities. *J. Ecol.*, **47**, 83–101.

WILLIAMS, W. T. and LAMBERT, J. M. (1960). Multivariate methods in plant ecology. II. The use of an electronic digital computer for association-analysis. *J. Ecol.*, **48**, 689–710.

WILLIAMS, W. T. and LAMBERT, J. M. (1961). Multivariate methods in plant ecology. III. Inverse association-analysis. *J. Ecol.*, **49**, 717–29.

WILLIAMS, W. T. and DALE, M. B. (1965). Fundamental Problems in numerical taxonomy. *Adv. Bot. Res.*, **2**, 35–68.

WILLIAMS, W. T., LAMBERT, J. M. and LANCE, G. N. (1966). Multivariate methods in plant ecology. V. Similarity analyses and information-analysis. *J. Ecol.*, **54**, 427–45.

WOODS, F. W. (1960). Biological antagonisms due to phytotoxic root exudates. *Bot. Rev.*, **26**, 546–69.

YARRANTON, G. A. and GREEN, W. G. E. (1966). The distributional pattern of crustose lichens on limestone cliffs at Rattlesnake Point, Ontario. *Bryologist*, **69**, 450–61.

YODA, K., KIRA, T., OGAWA, H. and HOZUMI, H. (1963). Self-thinning in over-crowded pure stands under cultivated and natural conditions. *J. Biol. Osaka Cy. Univ.*, **14**, 107–29.

ZINKE, P. J. (1962). The pattern of individual forest trees on soil properties. *Ecology*, **43**, 130–3.

Indexes

Author
Index

Subject Index